MW00562542

Navigating Graduate School and Beyond

A Career Guide for Graduate Students and a Must Read for Every Advisor

Sundar A. Christopher

American Geophysical Union

Published under the aegis of the AGU Books Board

Kenneth R. Minschwaner, Chair; Gray E. Bebout, Kenneth H. Brink, Jiasong Fang, Ralf R. Haese, Yonggang Liu, W. Berry Lyons, Laurent Montési, Nancy N. Rabalais, Todd C. Rasmussen, A. Surjalal Sharma, David E. Siskind, Rigobert Tibi, and Peter E. van Keken, members.

Library of Congress Cataloging-in-Publication Data

Christopher, Sundar Anand.
  Navigating graduate school and beyond : a career guide for graduate students and a must read for every advisor / Sundar A. Christopher.
      p. cm.
  Includes index.
  ISBN 978-0-87590-736-9 (alk. paper)
  1. Universities and colleges–United States–Graduate work. 2. Graduate students–Vocational guidance–United States. 3. Faculty advisors–United States. I. Title.
  LB2371.4C47 2011
  378.1'550973--dc23

                              2011043604

  ISBN: 978-0-87590-736-9

  Book doi: 10.1029/SP064

Copyright 2011 by the American Geophysical Union
2000 Florida Avenue, NW
Washington, DC 20009

Front cover: From seed to harvest, this image represents the enormous potential of the graduate student to succeed in every aspect of their career.

Figures, tables, and short excerpts may be reprinted in scientific books and journals if the source is properly cited.

Authorization to photocopy items for internal or personal use, or the internal or personal use of specific clients, is granted by the American Geophysical Union for libraries and other users registered with the Copyright Clearance Center (CCC). This consent does not extend to other kinds of copying, such as copying for creating new collective works or for resale. The reproduction of multiple copies and the use of full articles or the use of extracts, including figures and tables, for commercial purposes requires permission from the American Geophysical Union. geopress is an imprint of the American Geophysical Union.

Printed in the United States of America.

To
Sheba, Grace, Samuel, and Abigail Christopher
—true champions

CONTENTS

Preface

First, a huge thanks to the students who have taken the Professional Development course that I teach every year. And thanks to those who have provided valuable feedback. I really do enjoy the discussions, both in the classroom and outside. Next, special thanks to the people who have encouraged me to write this book (you know who you are). However, I would be remiss if I didn't mention Sue Glenn and Rita Sutton for believing in this even when I didn't! Special thanks to Kristi Caudill for diligently editing the manuscript. Finally, to my wife Sheba and my three precious kids, Grace, Samuel, and Abigail—thanks for believing in me!

Sundar A. Christopher
Huntsville, Alabama

1

Introduction

In the beginning...

Several years ago while I was a graduate student trying to finish my Ph.D., I knew that there was a real world beyond the exams and research. I was working on a dissertation and trying to finish a degree, but very few professors had time to advise their graduate students let alone mentor them in the next steps of their professional career. Oftentimes, the advising of graduate students was left to postdoctoral candidates (postdocs as they are affectionately called). The postdocs were also trying to navigate their careers sometimes without much success. Yet I saw some researchers and professors being successful, giving good lectures, presenting their work elegantly, and keeping all of their ducks in a row. They seemed to enjoy what they were doing.

At the same time, whether in academia or otherwise, I saw other professionals who were dejected and disillusioned with the whole system. They liked the research and work environments but hated the process—maybe because no one had ever provided guidance to them about how to maneuver a career path.

There were no courses or specific guidelines that taught graduate students how to navigate life in and beyond graduate school. Years later when I became an assistant professor, I was determined to develop a course that would help guide a graduate student through the numerous hurdles of graduate life and provide help for their upcoming career. I was and still am fully convinced that we need to provide students with the necessary tools to become successful in our respective disciplines and not lose them to disillusionment. Advisors have a professional obligation to empower students beyond just the classes and research.

Senior graduate students who are getting ready to defend their research have not even thought about writing a proposal. Neither do they know how to prepare and manage a budget for a project. They have very little information on what is expected of them when they graduate. Yet most students find that a place of employment requires them to communicate effectively, manage a myriad of projects, and write competitive proposals.

I distinctly remember several years ago, when I was a junior faculty member, apprehensively suggesting a course called "Professional Development." I wanted to teach the students the basics of proposal writing. The senior professors and the others in the room looked at me with glares that said, "Are you serious?" One even questioned why I would create more competition for myself since my livelihood also relied on writing and winning proposals. Finally, after some deliberation, they did approve a one-credit course that was to be taught only to Ph.D.-level graduate students. The course was initially centered on proposal writing and involved all the nuts and bolts of budget preparation, agency requirements, and templates and guidelines. The students loved the course and gave me some of the highest ratings. They felt that this was a skill that they needed.

They wrote an actual proposal for the NASA New Investigator program although they did not submit it. I walked them step by step through the entire process including writing letters of intent, drafting a 15 page proposal, reviewing the proposals of their peers, and understanding the selection process. I also discussed the bliss of getting a proposal accepted and how to develop a thick skin when one gets rejected.

Every time I teach this class I do get one common complaint: there is more work in this course than the one credit that they get at the end of the semester! Since then I have continued to talk with my colleagues in and outside my department, and I have listened carefully to the students regarding what they might like to learn in a course like this. Still called "Professional Development," the current course has a wide variety of topics including managing time and stress, presenting effective talks at conferences, and (get this—the most animated discussions I have ever had) managing your advisor!

I also have a steady stream of master's-level students wandering in to take this class. I continue to learn from these students and add topics of interest that will enable them to make a smoother transition from a graduate degree to employment. I also find myself giving more talks at student conferences these days, and the issues appear to be the same.

More than a dozen years have gone by, and I have been asked numerous times to put my thoughts down on paper. Well, here it is. This book is what I believe, and it is a sincere attempt to help the students navigate their graduate school lives and position themselves for success.

My colleagues may or may not agree with me when I say that most of us are too busy to do a proper job of advising students, let alone mentor them. I deliberately make a distinction between advising and mentoring as you will learn in this book. Mentoring takes effort, time, and discipline. This book, while a good read for students, will also be a useful one for advisors and faculty members both young and seasoned. While this book is not a substitute for proper advising and mentoring, it does take some of the mystery and haziness out of some topics that students always have questions about and might not get straight answers for.

I want to note upfront that it is next to impossible to capture the dynamics of a classroom when writing something like this. The discussions in the classroom continue to be an eye-opener for me, and I strive to include the topics and concerns that are important to the students. This one-credit class behind closed doors is something I look forward to every year. After a few "lectures," students fully warm up and won't stop talking. If instructors choose to use this as a text or reference of some sort, I have included some discussion questions at the end of each chapter to help lead discussions. This course was taught to a group of atmospheric scientists, and I am hesitant to state that every chapter in this book will be applicable to every discipline. However, I do believe that graduate students who embark on a journey to get a master's or a Ph.D. have some core set of issues that they deal with in their graduate life. I have tried to address these issues honestly.

I do realize that there seems to be a dichotomy that exists between students who only want a master's degree versus students who want a Ph.D. I strive to provide discussions that will help those on both of these paths. Regardless of which path one might take, most of these topics are highly relevant. After all, both are graduate degrees with high expectations from the student's future employer. I have tried to make the student "think" about what their future careers might demand of them so they can better prepare. This is one of the major focus areas of this book.

Finally, I have to say this about myself. I am optimistic by nature. I believe that you the student have an excellent seed within you called potential. Through hard work and the right preparation, you can succeed at whatever you put your mind to. I have seen students who, through sheer determination and hard work, outshine some who were "talented." Preparation and hard work seems to be the greatest equalizer. Another thing, I grew up in the American graduate school system and the way of competitive research, so a lot of the discussions will tend to lean that way.

So get a cup of coffee or boysenberry juice or whatever it is that you drink and enjoy the read!

I am hoping to hear from everyone who reads this book. So e-mail me or post notes for me. Let's get this started. . . .

Best wishes for a brilliant career!

Notes

2

Sowing

Cast a Vision

Without vision there is chaos.

One of the first things I ask the students in my Professional Development course is this: When should you start planning your career? After a few moments of silence, various responses start to crop up. Trust me. I have heard the gamut from "I will think about career planning before I defend my thesis/dissertation" all the way to "I have been planning my career since I was five." Admirable (or not) as that might be, most graduate students have given very little thought to career planning. While some may have thought about career-related issues, a deliberate plan has not been formulated to achieve those career goals. Many of us change our minds when it comes to careers. I should know. I used to be an engineer, but I am now an atmospheric scientist! However, when we get to graduate school, it is important to know where we would like to be after we graduate so we can formulate a plan of action while in graduate school—and do it while smiling!

I am not going to write about short-, middle-, and long-range career planning. Lots of self-help books already do that. What I would like to discuss are some simple steps that constitute career planning. I fully realize that your career aspirations can be very diverse. This book may not cover every single path, but the basic principles are the same.

I've said this several times during my course. You are in graduate school because you choose to be and because you are willing to work hard to achieve your goals and get to your destination. You had other choices when you finished your undergraduate degree, but you chose to come to graduate school either to learn more or to make more money. Maybe, if you are like me, you like everything about academia. Whatever the reason, you were willing to put in 3 to 5 years (I know some who have spent 10 years getting a Ph.D.!) and long hours to get this degree.

The myriad of classes, exams, presentations, projects, and papers is often grueling as you negotiate the good and the bad instructors, the competent and incompetent staff, the paperwork, and the all-encompassing bureaucracy. You try your best to maintain a decent semblance of a social life, but everything draws you back to one thing, "*I need to FINISH my degree.*" As human beings, we like closure and are propelled to final destinations. When we get there, we start driving again!

Given this backdrop, I have always maintained that you should think about where you want to be when you graduate when you start graduate school. If you already have a plan mapped out and know how to position yourself for success, then you don't need to read this book. Jokes aside, it is important to have a game plan for what you would like to do after you graduate. I say this because if you don't, you will never have focus during graduate school. *You become a wanderer*! If you know what you want to do afterward, then you can prepare accordingly in graduate school, fully geared toward that plan.

For those of you who have not done an iota of planning and wished you had, it is never too late to get back up on that horse and start riding it. I have had several students who postponed taking my class until just before graduation and have remorsefully remarked, "I wish I had taken this class earlier!" The major take-home point for these students is that they squandered their time in graduate school without focus and without learning how to position themselves for a successful career. They never stopped to think about what their future employer will be looking for and how to empower themselves for a successful career.

It is absolutely critical that in the first week of graduate school you buy a journal for your career planning. Write down notes and thoughts during your time in graduate school. This journal is not meant to be a daily entry, because I know that will drive you crazy. Write down your career plans, successes, and anything career related. Note that this will continue to evolve during your graduate career.

Later on, we will talk about analyzing your strengths and weaknesses. This too could be in your journal! Write down your highs and lows. Then when you hit a slump, you can look through your journal. With this in mind, here are some practical tips on career planning:

1. Resolve to think about career goals and how to prepare for employment from the very first week of graduate school.
2. Allocate at least 1 hour per week in the first semester of graduate school to think purposely about what those career goals might be.
3. Map out actions that you could take in graduate school that will help you achieve those goals. This is not transcendental meditation or useless chanting—it is purpose-driven steps on how to plan for your career.
4. Write your vision, goals, and plans down in a journal. This will be of tremendous use as you move toward graduation.
5. Talk to a peer who is highly focused on career goals to develop your own strategy.
6. Identify a mentor to discuss career goals on a regular basis.

As a student, you've probably seen some of your peers who are incredibly focused on classes and career goals, and you might wish you could buy some of that focus at a discount department store! The secret is simply one thing. These students know their destination, and they drive relentlessly toward their goal. They make sure that everything they do in graduate school will help them achieve that goal.

Let me whet your appetite by giving you an example. Let's assume that you want to be a university professor after you graduate (come now, it's not that bad of a job!) or you are willing to postdoc (more on postdocs later) for a couple of years before landing that dream job of yours at a prestigious university. Let's assume that this is the marker that you are driving toward. Do you know how you can prepare for this goal while you are in graduate school? You have to look in the mirror and see yourself 5 years from now and from your potential employer's perspective—gray hair and all! Your grades, your projects, your term papers count very little! Professors are judged by research, teaching, and service. Now if you can build your portfolio on these three items while in graduate school, you definitely have a heads up against the competition. This is what I will be talking about throughout the book—**sowing now to reap rewards later.** You've probably heard that famous, but very true, adage: What you sow is what you will reap! We'll be talking about this paradigm throughout this book.

Take-home message

 Begin planning your career when you enter graduate school and be purpose driven in graduate school. This will help you achieve your career goals. Spend at least 1 hour per week (preferably not during a sporting event) deliberately mapping out strategies to achieve goals. You will probably spend extra time on "career thinking and preparation" toward later stages of your graduate school. Much of graduate school is about positioning yourself to launch into a successful career.

Journal Entry

 Record the job you would like to have after you graduate. List the types of knowledge, skills, and abilities you think the job will require.

Food for thought

1. Where would you like to be after graduation? What type of employment will you seek? Once you have some thoughts, take some time to find out what skills that job will require (After all, I've suggested that you take 1 hour per week doing this—so now you have something to do!)
2. Without reading the rest of the book or looking at the list of topics, what do you think should be your career preparation steps?

Do a SWOT
(Strength, Weakness, Opportunity, and Threat Analysis)

Show yourself approved.

There are very few things in this book that come off as being too "textbookish." I promise that this is the only chapter that reads that way! I could have masked the whole thing by giving you veiled guidelines that smell like a SWOT analysis, but that wouldn't have helped. You would have figured it out anyway and then sent me one of those nasty e-mails.

What is SWOT? It stands for strength, weakness, opportunity, and threat analysis. Pretty neat, eh? All this means for you is that you should identify those four elements at periodic intervals in your graduate career (actually textbooks will say that you should do this until the day you die, but it's tiring when you get to be my age to be doing SWOT analysis all the time). It is good to honestly assess those four elements that you think might prevent you from finishing your degree, landing a job, or growing in your career. I'll ask you to do a SWOT analysis on various items in this book, and you'll find that it is indeed helpful—in spite of it being a bit textbookish!

So no pretense here! We'll go through SWOT analysis but with some practical tips. It is important to think about your strengths, weaknesses, opportunities, and threats to see how you can minimize and overcome weaknesses, maximize strengths, and mitigate threats. Why is a SWOT important? Evaluating your SWOT periodically while in graduate school and even after you land that dream job of yours gives you a better focus on how to prepare for the future. This is very much like your money. You've heard financial advisors talk about evaluating your portfolio periodically and readjusting things so that your target is achieved. While I hate to draw an analogy between career and financial planning, you have to agree that they are tied together reasonably well.

Here is an example. If you hate a certain class (let's just say fluid dynamics for an example, one of my favorites—yeah right!) and if it is indeed required for you to get over the hurdles to reach a goal, then you have to find a way to minimize or overcome that weakness. We all revel in our strengths, and we tend to do things to maximize them. This is not bad in itself, but it is like trying to strengthen one set of muscles while not paying attention to the weaker ones! Sooner or later something gets out of kilter. Take a look at some of those "out of kilter" jocks coming out of a fitness center—the ones with buffed up and weird looking out of proportion bodies! You need to keep your career muscles in proportion. Keep working on your strengths but work on the weak areas of your career as well.

I may draw some criticism for this remark, but I find that a lot of students who come to graduate school are exceedingly afraid of acquiring solid programming skills. Their reluctance to learn programming is so overwhelming that they go out of their way to put together a program of study that does not involve programming classes. Trust me when I say this—graduate school is the best place to put yourself in uncomfortable situations and learn new things. Only if you flex your muscles, can you begin to build them. A solid programming background will boost your self-confidence, because you will have the strength to tackle programming issues. Programming is usually logical (note I use the word usually) and helps in developing analytical skills. Take programming courses and other classes that require computer skills and put yourself through a tough semester or two. You'll thank me for this piece of advice!

Communication is another area of weakness that I have noticed in some students. For example, students are usually nervous about giving seminars. You should put yourself in situations that require you to communicate. Later, I will provide some practical tips for communicating better.

With that in mind it is quite useful for you to go through a career development exercise. It is critical to have a clear understanding of your career objectives and the skills required to attain them. Research shows that many new entrants into the workforce are poorly informed about careers and how to advance in their careers. A dean of a certain mid-sized university remarked that he was having a tough time motivating his faculty to aspire to new heights. Read that again—his faculty was having a hard time! In addition, research shows that careers are becoming increasingly self-managed, meaning that it is more important than ever for students to understand what they can do to develop their own careers. With a myriad of e-tools available these days, more students are able to be in the driver's seat for managing their graduate school and careers. That's good news!

Take-home exercise (no programming required for this one!)

 Your task is to write a personal career development plan. Your essay should answer the following three questions:

1. Where do I want to be?
2. Where am I now?
3. How am I going to get there?

A two- to four-page summary of your SWOT analysis will be a highly useful tool for your journal.

Here are some guidelines for each question:

1. Where do I want to be?
   - List your short- and long-term career objectives.
   - Determine what kind of career progression you expect to follow (linear, expert, spiral, or transitory)? See notes below for career progression paths.
   - Establish what you are looking for in an initial postgraduation job.

If you've never heard about career progressions, here is a short summary:

Linear—A progression of movement up an organizational hierarchy to positions of greater responsibility and authority; motivated by desire for power and achievement; variable time line; the traditional view of a career in the United States.

Expert—A devotion to an occupation; focus on building knowledge and skill in the specialty; little upward movement in a traditional hierarchy, more from apprentice to master; motivated by desire for competence and stability.

Spiral—A lifelong progression of periodic (7 to 10 years) moves across related occupations, disciplines, or specialties; sufficient time to achieve a high level of competence in a given area before moving on; motives include creativity and personal growth.

Transitory—A progression of frequent (3 to 5 years) moves across different or unrelated jobs or fields; untraditional; motives include variety and independence.

2. Where am I now?
   - Complete a personal career-related SWOT analysis: List your strengths, weaknesses, opportunities, and threats (i.e., those things outside your control that might keep you from pursuing your career plans and succeeding in graduate school).

3. How am I going to get there?
   - Outline a course of action to help you obtain your goals in light of your current situation.
   - Focus on one or two areas that you want to develop. One option might be to consider ways to leverage your strengths and to build upon them in order to exploit your opportunities. Another option might be to consider ways to address your weaknesses to help you to overcome them.

• List specific action steps you can take to help you attain your goals.

Journal Entry

List your strengths, weakness, opportunities, and threats on four separate pages. Make sure that you mark the date since these change with time.

Food for thought

1. Is a SWOT analysis really necessary?
2. Do you tend to only work on your strengths and avoid weak areas? How could this be detrimental to your future career?
3. In an ideal world that has no threats, what type of career will you pursue? Will your perspective change in the real world you live in?
4. What are some threats and opportunities in graduate school? How do these compare with threats and opportunities in your future job?

Avoid the Resume

Write down your vision.

Alright now that I have your attention, I have to confess that I was slightly overstating the point. You really do need a curriculum vitae (CV) or resume because every job announcement asks for one. There are lots of books about writing a CV and resume. I strongly suggest that you check out Peter Fiske's book, *Put Your Science to Work*. He discusses at length the differences between a CV and a resume and provides examples on how to go about writing one. He also talks about cover letters. I will only provide some brief thoughts.

What is the major difference between a resume and a CV? A resume is short (one to two pages) and is often written by students seeking a job at the bachelor's or master's level. A CV, on the other hand, is long. It provides extensive information on career goals, objectives, and achievements.

My whole premise is that if you have done the right things in school, your CV or resume is merely a placeholder or a requirement. You are already the leading candidate for the job and you have all the skills, so why bother with the CV? Brash as it may sound, it's the truth! You are probably still unsure of this method, so let me explain. In a later chapter, I will explain how you can ditch your dissertation by writing papers and seaming them together, but right now let's talk about how to pay a little less attention to a CV.

If you are applying for a job where you have no idea what your chances might be, then every word counts in a CV/resume, and you have to go through a lot of pain to get it to look just right. On the other hand, if you have done all the correct things in graduate school, and I hope you will after reading this book, then the employer is looking for you and not the other way around. Your track record while in graduate school should speak to the potential employer.

Then what makes you or your CV attractive? There are several things that are important to a potential employer. If it is an academic enterprise, then they are looking for certain criteria. They want someone with a good publication record indicating that you can write and communicate well and proposal-writing experience providing the employer with confidence that you can attract extramural funding. It is also important to have good communication skills (see section on how to present), teaching experience (working with advisor on class lectures), and service history (committees, networking, etc.). So if you know what your potential employer is looking for and you work toward that WHILE in graduate school, then you become the leading candidate for that job. If you fulfill some or all of those metrics, then the employer wants you. You will have several job offers. Therefore, your CV writing and agonizing can be avoided in terms of what font size to use and how to make it look perfect. This does not mean that the CV should be sloppily prepared—it simply means that the actual CV in its paper or e-form is a nonfactor. (See Web resources (http://sachristopher.wordpress.com/) for samples of CVs and resumes.)

Bottom line: Know where you want to work and determine what that future job entails. Then make sure that in graduate school you build those assets so your future employer has little choice but to hire you. Sow now to reap later!

Notes

3

Your Advisor and You

What Your Advisor Expects From You

Give honor where honor is due.

This chapter should be worth its weight in gold because I have not seen or heard an upfront discussion on this topic anywhere. This may not be exhaustive by any means; nothing really is, but I plan on writing other books!

For the purposes of this book, I am going to assume that you are a master's or a Ph.D. student and that you are on some sort of a graduate research assistantship that pays you handsomely (!) to do research, grade your advisor's course exams, and answer questions from the students in their class. Most advisors won't admit this, but they really want each one of their students to finish their Ph.D.s so that they can hire them. OK, that was a gross oversimplification. However, it is a reproduction principle. The advisor wants to graduate a student who can be successful and independent and who enhances their reputation and that of the university. Tall order, you might say. Not really!

I often surprise my students by asking them when they think they will be ready to graduate. I get a variety of responses. Some are smart-alecky: "When I finish my dissertation and have defended." Some are more somber: "When I have made my advisor happy." Be that as it may, here is another oversimplification that I throw at my students a lot: You are ready to graduate when you know more about the research topic than your advisor. It sounds simple, but sometimes it is hard to place a metric on that statement. Remember that the advisor sees the growth of a student from day one and is probably one of the best judges of the scientific capabilities of the student.

So what does the advisor expect from you? You may have thought that if you knew the answer to that question, then you could do things better. Well, wait no longer. Here is what your advisor expects.

First, your advisor expects you to pass all of your classes. They do not like to see mediocre grades, and they do like to see excellence. They definitely do not want to hear from another faculty member how poorly you are performing, how your work is sloppy, or how you show up late for classes and even for exams! Yes, I had a student who walked in to take an exam 15 minutes before closing time for an 80 minute exam. Time management issues, would you say? Does it surprise you when I say that faculty members discuss student performance and attitudes? Yes they do! What do you think they do at faculty meetings or in faculty lounges? Advisors want their students to do well. This doesn't mean that the advisor expects straight As from all classes (although that's nice!) but simply a solid understanding and an earnest effort to master the fundamentals.

Next, the advisor wants you to show initiative and take ownership of the research projects. Ownership is critical on all counts, but more on that later.

Most advisors expect their assistants to be in their office during posted hours. Here is a serious piece of advice if you want to stay on top of impressing your advisor and managing time well. Put your office hours on your door and send your advisor an e-mail so they will know when you are there. Your advisor may be the kind that silently watches your office! That's poor management practice on the advisor's part, but I'm surprised as to how many advisors actually keep an eye on their students' working hours.

The advisor wants you to be proactive in developing a program of study. This is your career; therefore, you should be thinking more about what classes to take rather than having them decide for you. However, ask for advice and pay attention because it is more than likely that your advisor has seen many students through.

Your advisor wants you to initiate meetings, make contact, and show up on time for appointments. After all, they have more on their plate than just you. They want you to show up to the meetings on time and have a plan of action.

Your advisor wants you to pass your qualifying/preliminary exams without much fuss. We all know how hard these exams can be, and whining and complaining about format and exam questions really does not help. Get over the hurdle and keep your focus on your career.

Your advisor wants you to take ownership of the project you are working on. This gives the advisor much more confidence in the outcome. Imagine that you take the initiative to create a blog or Web page and list all the papers relevant to your research with your comments. I bet your advisor would be beaming with joy! Advisors grow tired when students repeatedly ask where to find resources rather than being resource centers themselves. I believe without a shadow of a doubt that every student working on one or more research topics should become a resource center. More on that later.

Your advisor will be wildly excited if you take the initiative for writing your first paper on your research results. I have yet to see an advisor who has turned down the first draft of a paper written by a student for journal submission. (Speaking of writing, I'll provide more tips on how to completely avoid writing a dissertation later. I could get in trouble for this advice!)

Volunteer to give talks, either in your group or in your department. This tells the advisor that you are willing to mature.

Never ever show up for meetings with your advisor with body language that says I'd rather be fishing. Worse still—do not give your advisor results and graphics that were just printed out a few minutes ago. It is better to simply reschedule your meeting (don't overdo this) rather than provide a halfhearted attempt at discussing your research results. Notice I said "your research results" not your advisor's.

Stay proactive in graduate school and take initiative in your research. This goes a long way toward building a healthy student-advisor relationship.

The list of proactive items is endless, so be creative.

Bottom line: Become a proactive student!

Journal Entry

List 10 ideas for staying proactive that will enable you to be a better graduate student. Send me those ten things via e-mail, so I can list them in my next edition!

Food for thought

1. Reverse your roles for a minute. You are the advisor. What would you want your student to be and do?
2. Switch back to reality. What are the weak areas where you need to be proactive? How can you fix them?
3. Are you proactively managing graduate school? If so, how, and if not, why?

Manage Your Advisor

Isn't that a thought?

I almost wrote a huge preface to this chapter because no other discussion topic has ever brought a spark to the class discussion more than this one: the advisor-student relationship! But I decided to take a different approach and actually made this a proactive topic—**manage your advisor!**

Picking the correct university for graduate school is no easy task. Some students want the reputation of a "top three" university, even though the specific department that you will be in may be way down the totem pole in ranking. This is a big mistake if you are into things such as rankings. Others want to go to a university that has a beach, a good football team, a small hometown feel, a cool summer—the list is endless. For some, the research assistantships and their stipend are important. I had a student who noted that by entering graduate school the opportunities for making money in the real world are compromised until graduation. Whatever your peripheral criteria might be, make sure that it is a good academic program that provides both the breadth and depth of knowledge and experience that is necessary for you to succeed. Check out the university's course offerings, faculty, students, and available research portfolios.

Often the most forgotten and poorly thought out piece in this mix of metrics is the advisor. This should be one of the most important criteria in your selection of a university and a program—especially if you are going to be a research assistant. Why so? Because you will spend at least 3 to 5 years of your life with this man or woman who will also go through mood swings (much like yours), who will face different pressures than you do, and who may sometimes be seen walking the corridors muttering underneath their breath. If you are one of my students, you may sometimes hear a huge—and I mean a HUGE—sigh from my office with the accompanying statement, "I can't believe this stupid bureaucracy!" Hey, confession is good for the soul, right? If you want to have a good student-advisor relationship, then start thinking about your advisor's perspective. This means learning how to walk in their shoes. Granted, this is no easy task. But if you can learn to think from your advisor's vantage point, things get a bit easier.

Advisors are not in the same phase of their careers as you. If you know the stress factors in their life, you can navigate the student-advisor relationships effectively—or at least reasonably well. As one of my students put it, learning how to "manage" your advisor in graduate school is a good learning experience for future scenarios that may include good and/or bad supervisors.

When I ask my students about their expectations for their advisor, I get such a talkative class with a list so huge that I have to chuckle. I make sure that I write down their entire wish list on the white board so they can see it. The expectations are huge. I am also absolutely sure that the expectations of student virtues from an advisor vantage point will be daunting. More later on that!

In short, most students want to be able to interact with their advisor on a near-demand basis. Students want complete and full access to the advisor at any time and comprehensive guidance on research and class affairs. To top that, students also wish for the advisor to be interested in their careers and well-being. Some students require high-quality nurturing time and want their advisors to be their friends. If you look through this list of needs and wants, you will quickly see that students expect advisors to be more than a guide for research and academics. They want a mentor!

I don't know the dictionary definition and contrast between an advisor and a mentor, but I tell my students that an advisor guides the student in research and academic affairs only and that a mentor does all of that but also takes responsibility for shaping the career of the student. The mentor has the student's career in mind and is genuinely interested in passing on the baton. The mentor's job is not done even after the student graduates. They continue to take an interest in the post-graduation life as well. Much like parenting, the mentor never fails to keep up with the student's accomplishments even after graduation and provides targeted advice when necessary.

I shock my students when I state that not every advisor is equipped to be a mentor. Therefore, expecting mentorship from your advisor may be unrealistic. *This expectation mismatch is what provides for some unhealthy tension between the student and the advisor.* Keep in mind that not all advisors want the responsibilities of being a mentor. My simple solution is this: If your advisor is not the mentoring type or does not want such responsibilities, then there is nothing wrong in seeking mentorship from others, preferably in your department or at least in your disciplinary area. A mentor is important. Period. Keep your advisor but seek out a mentor for filling the gaps.

It is easy to spot a mentor. Their door is usually open, and they take a genuine interest in nurturing students. They are really interested in seeing their students succeed. Students tend to migrate to their office to discuss research, career, and other issues. The mentors are genuinely interested in the next generation and do not pay lip service to student affairs. They have the students in mind.

In the graduate career, most—if not all—students go through setbacks, disappointments, and the occasional "grad school blues." If you expect your advisor to get you out of that slump, and they do not have an idea how, or they don't have the time to do this, you are setting yourself up for some major disappointment. Oftentimes, crying on a peer's shoulder is not a good either since they have no idea what to say or how to react. A family member or a neighbor also cannot relate to these grad school blues if they have not been through that process themselves. Therefore, finding someone who can serve as a solid sounding board and who can provide balanced advice is necessary. I still knock on the door of my mentor when I get stuck in situations, and he always provides a balanced perspective and helps me navigate the challenge. Priceless!

Most of you who are reading this book already have advisors, so there is no point in telling you how to pick one. (You may be dying to hear my viewpoint, but you will have to wait until later in the book to find out.) If you have a research or academic advisor, the first thing for you to realize is that they too have pressures and stress factors in their lives. Try to understand these stress factors and learn to navigate them well. This will help you have a healthy relationship with your advisor. Know your advisor, their expectations, and their work patterns.

Journal Entry

 List the strengths and weakness of your advisor when it comes to managing students. How can you as a student maximize your advisor's strengths and mitigate some of their weaknesses? Here are some practical guidelines:

- **Find out the working rhythms of your advisor**. Do not expect to walk in or demand appointments on days when their work loads are usually higher. Some advisors teach on certain days and do not like to advise students on those days. Know what these patterns are and respond accordingly.
- **Take initiative**. Advisors love initiative. I certainly do! When one of my students goes above and beyond what's required of them, it tells me that they have a drive to succeed.
- **Be proactive**. Do not wait for your advisor to draw up a program of study or suggest writing a paper. Remember that it is your career that is at stake.
- **Initiate meetings**. Meet with your advisor regularly. Keep in touch. Remember that your advisor may be juggling several projects, classes, travel, and other things in life. Therefore, staying in touch with your advisor is your responsibility—not theirs! Trust me when I say this: an advisor who does not initiate meetings is still looking for productivity. They may be called "hands-off" advisors, but they expect the same level of productivity (if not more) as a "hands-on" fully engaged advisor.
- **Do not wait for your advisor to send you e-mails/advisors**. Remember that most advisors are busy. When and if they do send you an e-mail, respond as soon as possible. Do not procrastinate!
- **Do not ramble at one-on-one meetings**. Time is precious. Have a list of things to talk about at each meeting with action items. Follow up with e-mails.

- **Show results.** At meetings with your advisor, be prepared to show your progress through your graduate career.
- **Remember that there is no such thing as a perfect advisor or even a perfect mentor.** If they have not offended you yet, they will soon. Learn how to think from their perspective. In other words, try to look at things from their vantage point!
- **Learn how to manage your advisor.** This is valuable experience for your future career when you have to manage your future supervisor or team lead.
- **Learn how to draw from your advisor's research and other career experiences**. Remember that they have also gone through graduate school, and they have overcome similar obstacles and stresses. They have a wealth of research experience and a proven pattern for success. Your job is to draw from their experience. Learn all you can from both research and nonresearch issues. If they are good at multitasking and managing time, observe and emulate. If they are good communicators, learn and emulate. The list is endless!

Take-home message

Your graduate school experience and your career are your responsibilities and not your advisors. Stay proactive.

Food for thought

Do a SWOT analysis on your advisor based on what you know. Wait, DO NOT e-mail that to them.

1. What are five of their strengths, weaknesses, opportunities, and threats that prevent them from being a better advisor or mentor?
2. What are some concrete steps you can take to minimize their weaknesses through effective communication? (Example: Maybe their weakness is project management, and it creates stress in your relationship with them. How can you mitigate this?).
3. What can you learn from their strengths as an advisor that can benefit your graduate career?

4. Not all student-advisor relationships are conflict-free. What are five top items that produce tension? As the student, what can you do to minimize and maybe completely negate potentially explosive situations?
5. Try role reversal. If you were the advisor, what types of opportunities would you provide for your graduate student to learn and mature?

Notes

Skills

Work Ethics for a Graduate Student

Work not for your advisor but for a higher purpose.

Times have changed. You have probably either heard your professors voice that openly or you have overheard them in a conversation on your way to class. Maybe you lingered just long enough (note that I did not say eavesdrop) to find out that the professor was talking about the work ethics of a graduate student. You shook your head as you walked away thinking, "I used to hear my Dad say that back in the good old days. He climbed a hill barefoot—both ways—with mountain lions chasing him." Jokes aside, there is a sense that work ethics have changed in younger generations. They just don't seem to work as hard or read as many papers or code as many lines or work on weekends.

Some students do have poor work ethics. Notice I said "some." Those students never show up on time, have a rebellious attitude toward their advisor and the graduate school system, and constantly question productivity and excellence metrics. But in my opinion, that is not the norm. Students face different types of social and technological pressures than their professors faced. Therefore, the work ethics seem different but not necessarily poor. I am going to guess that the students reading this section like what I am saying but that the advisors are mumbling. Remember though, every student will become an advisor in some capacity at a later date, and every advisor was once a student. Sobering!

I teach a demanding class called Satellite Remote Sensing that requires a lot of programming on top of projects and tough assignments. It is an optional class. The students who take it either want to learn the material or their advisor forced them into it. Others avoid it like the plague! The students affectionately call it SARS for Satellite Remote Sensing, but SARS also stands for a deadly virus!

A few years ago, I had a male student who was extremely sloppy. He finished only about 30–40% of each assignment, came late to class, and struggled with the course. I found out later that the only reason he was there was because his advisor thought that taking my class would bring out some academic discipline in him. I hate it when advisors send their problem students to my class! As always, I made it a point to write in his papers and assignments and let him know that, at this rate, he would not even pass the course. Considering that this is a graduate-level course, a failing grade has serious implications. One day, after about 6 to 7 weeks had passed, he stormed into my office and rudely told me the following: "Your class leaves no time for my social life, and I am having a difficult time getting along with my girlfriend!" He continued his tirade for a few more minutes, and I still did not respond. At the end of his tantrum, I looked at him and asked him if he wanted to rephrase any of that as a question so that I could respond. He stormed out of my office the same way he came! Needless to say, he ended the class with a grade of D.

I recount the story to say this: poor work ethics could ruin your graduate experience. Students like the one I just mentioned are plenty smart, but they do not want to invest time and energy into getting a good graduate education. My question is this: Why put yourself through the graduate school experience if you are not willing to work hard?

For every one of these negative experiences, I can recount several positive ones where students are genuinely interested in learning the material and moving ahead in their careers. A graduate student who is determined to excel has good work ethics and a desire to master good work habits. I'll tie this to ownership. If a student has taken ownership of their research and is in a forward thinking mode, then good work ethics and habits are not a problem. If they have no concept of ownership, then it is an uphill battle for both the student and the advisor.

For instance, one professor had a very difficult time gaining any sort of productivity from his student. Since the professor always came early to work, he told his student that she must come and get her office keys from him every morning at 8:00 A.M. He thought that this would promote accountability, but it only created more friction between the student and the advisor. The advisor should have focused on building a relationship of trust to help the student take ownership.

I should let graduate students in on a secret. The advisor is usually fully aware of the amount of time it takes for an assigned specific task—give or take a few hours. Since the advisor came through the student ranks at one time, they *should* be good at this estimation. So when a student shows up for a meeting and makes a 3 hour job sound like it takes 3 weeks, it does not promote a relationship of trust. Neither does it signal to the advisor that ownership transfer is underway. I am often amused when a student walks into my office for a meeting and proceeds to tell me that it took them an entire week to analyze a piece of data and make a graphic when I know full well that, at most, it would have taken a day or less. This quickly breaks trust between the student and the advisor. These situations put the advisor in a position where they have to enforce measures to monitor time rather than progress, which breaks down the mentoring process. This is indeed a lose-lose situation for both the student and the advisor.

As a student, you must seek independence and ownership. Don't force your advisor to micromanage your office hours. It is hopeless for you and the advisor.

Time management is one of those areas where I am asked how I manage my students. As a general rule, I do not micromanage any of my students' time. However, I do ask them to post the hours when they will be in their office. If I need to see them to convey an important message or concept, I will know when to find them. I am a walker by nature. I have to get out of my office and interact with my students and even with other students whom I do not advise. However, I never ever go to students' offices to check to see if they are present during office hours or not. Never! I gauge students' productivity by their growth in ownership, their progress, and their initiative, enthusiasm, and willingness to take on tasks. I preach teamwork at all times, so I do not look for superstars in my group.

As a student, your work ethic should speak volumes not only to your advisor but also to your peers, other faculty, and other members of the department. When in doubt, go that extra mile with your work to provide the best quality of work that you possibly can. Graduate school is absolutely a great training ground for your future job, and good time management is a skill that your future employer expects.

A relationship of trust between advisor and student is vital for productivity. Don't break that trust with poor work ethics. Communicate with your advisor on what your work patterns are and ask for advice on whether you need to fine tune them to meet expectations. Let your advisor know what your work hours are and when you plan vacations and time off.

One of my students came up with the brilliant idea of maintaining an online calendar and giving me access to his schedule. Now I know exactly when he will take his exams, when the first draft of his paper will hit my desk, and when he will be on vacation. Everything! I call that taking initiative. I love it! So create an online calendar and share this calendar with your advisor. Google Calendar is an excellent tool for this. Indicate your work hours, deadlines, and meeting dates. You should also indicate when your advisor can expect drafts of papers. Make sure you meet these deadlines. Being proactive with work ethic measures is the best way to let your advisor know that you are serious about graduate school and your research. Plus, if your advisor is not a well-organized individual, he could learn from you!

Journal Entry

 List some of your work ethics you feel are being noticed and some that are not. Is this a source of frustration?

Food for thought

1. What are some ways that you can avoid breaking trust with your advisor?
2. What ideas do you have for what constitutes good work ethics?
3. What are your expectations for your advisor's work ethics?

Take Initiative: Becoming a Resource Center

Without vision there is chaos.

I use the phrase "take initiative and be a resource center" in many instances in this book. But, taking initiative does not necessarily mean that you are overly aggressive and step on everyone's shoes, sandals, or socks! Let me give you some examples.

If you know your end goals well, then taking initiative is easy. One of my favorite examples involves a student who came from the armed forces to the university where I was employed. He was mature and willing to work hard. He imposed on himself a 4 year deadline to finish his degree, and he knew exactly what he wanted. Naturally, he had to take initiatives to drive the system that he put together. He did not wait for his advisor to tell him to get his program of study done. He did not wait for the advisor to suggest forming a committee. He did not wait to write journal papers. No one had to tell him when to take his Ph.D. exams. Everything was planned, submitted for discussion, and adjusted. He was constantly in the driver's seat. Students like this do not wait for research or programming problems to linger too long. They make sure that they either solve them or find ways to solve them with appropriate advice. He was not even a student of mine, but from the time he took my Professional Development class, he made it a point to stop by every so often to glean from my experiences and to seek advice where necessary. This is what I mean by taking initiative: being in the driver's seat! If you don't drive, then someone else will. Chances are that they will take you to the wrong destination. Plus, they may be a bad driver!

There are lots of ways that students can take initiative and become a resource center. Create a blog or a website with information about the research topic you are working on. This shows plenty of initiative. Do you remember the advice about reading papers and making notes? In this interconnected world, there are excellent tools to help in doing this. You can create a weblog, post your comments online, and become a resource center for your team and/or advisor. In fact, one of my students started a weblog on one of his areas of interest. Every time he read a recently published paper or something that he thought was a breakthrough, he quickly posted the title, an appropriate figure, and his thoughts about the paper. Many people now visit this blog, and the worldwide community provides valuable comments. This is indeed a classic example of becoming a resource center. You can also create a page for your team and your advisor. This will allow them to see your progress and the research papers that you have read. Again, remember that this helps build those communication muscles while showing initiative.

Apply for travel fellowships to go to conferences. You'd be surprised at how many there are around the world that you can attend. Prepare a poster based on your research and be ready to present at the conference or volunteer to give a talk at the next group meeting or seminar. If your goal is to become a professor in the future, then you should volunteer to give a few lectures when your advisor teaches. You can impress your advisor by developing course notes for that section. You can also use this when you actually teach a few years later.

As you mature in your research, you can also attend workshops—usually free of charge—taught by experts in your area of research during the summer. Look out for such opportunities. If you are not sure, ask your advisor or, better yet, surf the Web. There's plenty of information available.

Take initiative by writing a proposal to win a fellowship. It is a huge feather in your cap and a major dossier-building item, and it releases your Graduate Research Assistantship (GRA) for another student for your team. You could then mentor a young graduate student. A win-win situation!

Learn to get into the driver's seat in graduate school. You need to initiate all deadlines, paperwork, and paper and proposal writing. You can have a complaining attitude about your advisor, the department, the university, and other things around you, or you can take the high road by being proactive. One of the biggest growth spurts that a student undergoes is when she suggests to her advisor a series of papers she will be writing based on her research. This signals to the advisor that taking ownership has fully happened and that the student is well on the way to gaining full independence. Notice I used the phrase "series of papers," not just one.

Food for thought

1. Carefully note who among your peers takes initiative in their research and graduate school activities. Talk to them about what specific initiatives they take.
2. Talk to your advisor about how you can take more initiative in your research.
3. List conference and workshop activities that will benefit your career that you can attend.

4. List your metrics for maturing as a graduate student. Come up with specific signposts for maturity for every phase of your graduate career.

Take Ownership

It's your future!

In all my years of being a faculty member and a mentor, nothing thrills me more than when I see students take ownership of the project they are given and bring new dimensions to it. I have to confess that I learn from my students as much as they learn from me. It is a rewarding experience! If you are an advisor reading this book, I am sure that you will agree!

Taking ownership of a research project takes time, and it is a fabulous interplay between advisor and student as to how that ownership handoff happens! Some of the junior faculty members struggle to give up ownership while fully realizing that it is the best solution. Other advisors try to dump the ownership too quickly, and the student struggles with the load and expectations. Still other advisors have absolutely no idea what ownership even means. If done properly and with time, it can be one of the most rewarding experiences for both the student and the advisor.

Let me explain. On day one, when a student walks into an advisor's office, the excitement is high. Most students are thinking about classes, exams, and fitting in. Research is somewhere in the rearview mirror. Trust me—the end point for most students is getting a good job! The regular meetings then start, and the students slowly begin to gain momentum in their research work. Most students still think of research projects as their advisor's research projects. Some have an idea as to how these research projects came about, while others have no clue. Students often think that they are paid to do a certain job and that their advisor must explain all facets of the research. In short, the student perceives that he or she is a hired hand. Nothing could be further from the truth, since these are research ideas that have quite a bit of leeway to explore ideas above and beyond the stated requirements in the proposal. Good research is what the funding agency wants. Sometimes the agency requires deliverables in terms of products, but oftentimes the metrics are quality and quantity of papers that are written as part of the proposed project. The advisor has the task of not only articulating that the student must take ownership of the project but must also be willing to give up ownership while providing solid guidance. After all, the proposal was won on the strength of the advisor's proposal and reputation.

I often tell my students that if they think of this as their advisor's project, they will only go so far and no further. They read papers for the sake of reading and analyze data and graphics for mundane reasons. The meetings with the advisors are listless with very little in the way of innovative ideas. The students are simply waiting for their advisor to tell them what to do next.

However, if the student truly believes that this is his project, then innovation flourishes and productivity increases. The purpose for his graduate career and beyond is more meaningful. This is what I call taking ownership!

Some of my students have come to graduate school with ownership in mind. They excel right from the start, require little supervision, and prefer having a lot of independence. As an advisor, I have been able to hand over the reins and see the productivity level multiply beyond my wildest imagination. One particular student of mine stands out. He had already published several papers, and when I thought that he would begin to wrap up his work, he clearly explained to me that he wanted to embark on a related research topic. He spent an additional year writing another series of papers. Needless to say, he has been much sought after since his graduation and works at a reputable institution now.

Some students require a lot of prodding to get to that ownership juncture. I make it quite plain that I desire for them to take ownership and to come up with innovative ideas to challenge themselves. It takes time, but with mentorship, it does happen!

> *Note to advisors*: In the process of taking ownership, the student is bound to make mistakes, and sometimes over-eagerness or overzealousness can run rampant. As an advisor, take the high road and provide levelheadedness without resorting to shooting down the student's ideas or being overly critical. Be patient!

Research projects

Most projects, though not all, are competitively won. This simply means that the advisor comes up with an idea and puts it down on paper. That is called a proposal. The advisor submits this to the requesting agency, and the proposal is reviewed nationally, and sometimes internationally, by several experts. Out of a hundred proposals like this, 10 to 15 may be awarded with support for graduate students (stipend and tuition). The student gets a Graduate Research Assistantship (GRA) and works with the advisor on this project to fulfill the stated objectives in the proposal.

Take ownership of your projects. Be genuinely excited about them and learn how to innovate and come up with new ideas for papers. When an advisor is not willing to relinquish ownership as fast as you would like him to, then it is your turn to be patient.

Journal Entry

 When did you take ownership of your projects? Did the relationship between you and your advisor change when that happened?

Food for thought

1. Every advisor and team is uniquely different. In the research environment that you are working, can you list some of the things that you could do take ownership of a project?
2. Assuming that you have taken ownership of the projects that you are working on, what do you think are some metrics for success?
3. What are some of the challenges for a student taking ownership of a project?

Work Hard and Smart

Work hard, be diligent.

Through the years that I have been teaching this course, if I were to stop and ask any student in the hallway what they take home from this class, it is probably not going to vary a whole lot. You would probably hear phrases such as "taking ownership and initiative," but you would likely hear this statement the most: "Write your peer-reviewed papers!" I say this so much that I have been jokingly accused of brainwashing the students! I say "write your papers" for several reasons: it provides focus for research, it has a definable end point, and it is a golden metric for success. I am absolutely sure that some will disagree with such an oversimplification, but I have seen many students without focus in graduate school quickly become motivated when writing papers. This is because they have an end point to drive to and build on. Writing several papers during your graduate career will only help, never hurt.

Recently, I sat down with the head of a selection committee from a midwestern university in the United States who was responsible for interviewing and guiding the selection process for a tenured-track faculty member. He noted that the winning candidate had already published several papers even before his graduation. This seemed to be the edge that the department was looking for. Granted, this may not be case for every scenario, but publishing papers in reputable peer-reviewed journals signals that you have a high level of potential for most jobs. Yes, I know those who will argue that not all jobs require a publication mantle. Be that as it may be, it will never hurt your dossier. A curriculum vitae with a good number of papers signals to the potential employer that you have the capability not only to think through problems but to actually complete them. That is a valuable asset! It does not matter how many great ideas you have regarding research. Unless they are written down and published in a reputable peer-reviewed journal, they will not count.

Undoubtedly, a graduate degree will require good writing skills. Writing peer-reviewed papers even before you graduate will help solve multiple problems at the same time. Let's assume that your end goal is writing several papers before you graduate, and let's assume that five papers is a reasonable limit. (We'll talk about quality versus quantity issues later.) First, I want to convince you that five papers for a graduate student with at least three to four as the lead author is not out of the ordinary, and neither is it a tall order. So with that as the end goal, where do you start?

Your first task as a graduate student is to think of yourself as a resource center for the projects that you are handling. When you first start your graduate career, your advisor may (and should) give you a list of recommended reading, most of which will be peer-reviewed papers but some books will also be included. Over the course of your graduate career, you must be in the habit of reading and assimilating information from at least 10 papers a week. As you progress, you should begin e-mailing your advisor about new things happening in your field and the new papers that are coming out. This will indicate to your advisor that you are serious about keeping up with the literature. Sadly, many students, even during the finishing stages of their degree, have no idea what the relevant papers in their field are—let alone know how to provide a top-notch literature survey to the committee.

Knowing where the online journals and resources are is a first priority when you enter graduate school. You may start by reading five papers every week or two, but you have to end up reading at least two to three times that number before you finish your degree. The first thing students say to me is, "You can't be serious." My response is this: "You bet I am!" Look at the number of journal papers that stream out every week. If you do not come up with a way to keep up with the state-of-the-art research when you are a graduate student, you might as well forget it when you start your job. Put a system in place that will enable you to organize the papers you read. An online system (e.g., EndNote) or your own that you can access from anywhere is ideal.

The next thing that I am often told is this, "You cannot possibly expect me to read every single word of every single paper and understand every concept and equation." My answer is this: There are some classic review papers in your area of interest that you had better read every word of and make sure you fully and completely understand. Study those papers as you would any textbook. There are also other papers that you can read to glean important information. Nevertheless, it is a fantastic habit to read these papers and write some brief discussion down (half a page) of your thoughts about them. Better yet, put those thoughts on a Web page or a blog of some kind. There are a myriad of tools available these days to stay organized. Effectively gleaning information from papers comes with time, but the bottom line is this: read as much as you can and become a resource center. Again, the goal is to come up with a method to organize and assimilate information in a focused and deliberate manner!

Time allocation

In a world full of distractions, I find it best to allot time each week to go to the library or some other quiet place to read and study journal papers. Take a stack of papers, a pencil, and a highlighter. There should be no distractions, and this should be time well spent. Reading builds your research muscles. I still do this once in a while. In spite of the increasing pressures on my time, I tend to take a load of papers on my trips. Other times, I disappear to a local coffee shop or the library (far away from my desk!) to do some quality reading. There is always a stack of papers in my office that I want to read. If you are not sure in the beginning what to read, ask your advisors or senior students. They will be glad to help. It is indeed difficult to stay on top of all the literature these days, but if you don't try, then you will lose your competitive edge. Yes, I also rely on my graduate students to keep me current and to stay on top of the papers.

Practical tips

- Get memberships to the most common organizations in your field of study. For meteorologists and atmospheric scientists, there are many organizations, including the American Metrological Society (AMS) and the American Geophysical Union (AGU).
- Go to the websites for the most common journals where papers in your field of work appear. Most journals will send you a list of papers from each issue if you simply provide your e-mail. This is a great way to keep track of papers in your area.

Next, I will state this simply: All work (reading) and no play (coding/analyzing) makes you dull. For all the reading you can do, if you do not supplement it with actual coding and analyzing your ideas, then you can quickly get bored. Learning how to code, analyze, and interpret results effectively is an ever evolving process. It takes time. As long as you can get the job done without errors, you should be happy. You have read the papers related to your project, and you have your data set that you are working on, and you are beginning to make some impressive graphics. Never shortchange yourself by making poor graphics. Persevere for good labels and high-quality graphs and images for the projects that you are working on. Quickly, you will find that you are becoming a resource center for programming issues as well.

I probably have said it before, and I probably will say it once (or is it twice?) again: you have to put yourself in the frontline by purposing to become proficient in programming and using various tools effectively. In the sciences, it is imperative that we manipulate data and visualize them effectively to comprehend the results. There are a myriad of tools, and I am hesitant to recommend one, or even a few, because of the level of complexities involved. In the courses that I teach, I actually emphasize IDL—an interactive programming language that is quite powerful—for most things encountered by scientists and engineers. Whichever route you take, the bottom line is this: build your tool chest. There is nothing more frustrating for you as the student—or for your advisor—then when you are in the finishing stages of your graduate career and you are dealing with programming issues rather than concentrating your efforts on writing and analyzing your results. I strongly recommend that you take at least one rigorous programming class and other classes in your discipline that require programming. Although painful at first, they are well worth the rewards.

The next step is probably the most important: analyzing the results! This is where you understand the research problem and set the stage for your research. You, along with your advisor, will spend many an hour thinking about these issues. This is one area where I judge progress. The student, through independent thinking and armed with the papers that they have read, should progress steadily in analyzing these results. If your advisor asks for different types of graphics or further analysis, don't get upset. This is how research evolves. If you have not read the papers, analyzing your results could be bewildering. Prior research sets the context for any additional and new findings that you may bring forth in your research. Relating your results with previous literature is often the first step, and then you will begin to see new things in your data or your research to come up with new thoughts. Remember that most research evolves incrementally. There-fore, enduring discipline is needed while you are in graduate school. There are no shortcuts!

I strongly encourage the students to put each graphic in a blog or file followed by a short explanation of what you think the analysis might be. Given that the analysis evolves with time, it is a good and success-bearing habit to capture your thoughts. Plus, it gives you great practice at writing.

You must be proactive and discipline yourself to have good graduate student habits whether it is reading papers or analyzing results. Discipline will definitely reward you with success.

Journal Entry

 List what you would want your student to have for work ethics if you were the advisor.

Food for thought

1. List five items that you think are top-notch research habits.
2. Assess how much time you devote to each one of those five items.
3. Do a SWOT analysis on your research habits and formulate a concrete plan to assess these items periodically!

Be Whole

A divided house cannot stand strong.

I said earlier in this book that poor time management causes stress. That's not the only reason for stress these days. External factors could easily cause undue stress on a student. Poor health, family illness, and a plethora of other reasons can rob a student of the joys of a fun-filled graduate career. I want to note that I do not live in a vacuum and only teach the "publish and manage time" mantra even while there may be real-life issues at stake. However, it is my experience that very few students face external stress factors. When external stress factors happen, I recommend that the student take some time out to evaluate priorities and to make hard decisions about their graduate career. For the majority of others most stress is self-induced.

Graduate students seem to be getting better at handling stress. They seem to be getting on the bandwagon of eating healthy, exercising regularly, drinking lots of water, and sleeping enough hours in a day. (I also know of the occasional student who sleeps too much, misses classes, and shows up for exams late!) I find that the students who are on top of managing their graduate life well are able to stay on top of the "wholeness" concept as well. This is a great trend because these factors play a central role in the general physical, mental, and emotional well-being of a student.

As part of your time management portfolio, I strongly suggest that you draw up adequate time to have a balanced exercise regime. Several students in my class have mentioned that it provides significant mental agility to solve problems, and it pays off in the long run to eat well and exercise regularly. I couldn't agree more! For the younger students, strenuous exercise seems to be a great outlet, and most anyone can develop a sensible regimen of common sense exercise. Try racquetball for fun and exercise. You can beat a small blue ball as hard as you can (try not to visualize your advisor now!) Most campuses have a good biking system, and several university cities or towns have a well-laid-out bike path. Most universities offer free memberships to students at the fitness centers. I strongly recommend using these facilities. I know of a student who makes sure to get outside for lunch for at least 30 minutes each day and walk around campus to "energize." I am trying to do that now!

A note about social bonding! Student relationships built during graduate school could last a lifetime. Remember that the leaders in the research community over the next 30 years could come from your peer group that you are taking classes with and meeting at conferences. Get to know your peer group and build relationships. I remember when I was a graduate student, we did a lot of things together that I reminisce about even today. But enough nostalgia!

Unwinding is a necessary part of graduate student life. While some crave a high-octane exercise regimen that revs up the endorphins, others make it a point to commune with nature during the lunch hour. I have known other students who love to cook. Whatever unwinds you and provides mental relaxation is important to do. One of the best pieces of advice I got as a graduate student from a well-known professor was to disappear for a set period of time (a few hours to an entire day) every so often without telling anyone of my whereabouts to do some creative thinking. This place could be anywhere from your study in your home (I do not recommend this for everyone due to obvious distractions) to coffee shops or hiking trails. Hold on to that passed-on advice. You'll need it in the days to come! Where do you think I am writing the final stages of this book?

Learn how to identify stress factors in your life. Before it gets to you, take measures to avoid the effects of stress. Manage time well and, of course, eat well, exercise often, and live long! Productivity is definitely linked to the well-being of the whole you!

Again, here is a personal note since I've been asked this numerous times. What do I do? Exercise is critical in my life. I play racquetball at least three times a week and, I consume lots of water, yogurt, and fresh fruit. I hide for at least 4 to 6 hours per week in my favorite place to put in some "creative think time" and writing. If it does not happen every week, I accept it and move on to the next week.

Food for thought

1. What are some of the external stress factors in your life? How can you realistically manage those stress factors?
2. If you have not paid attention to the exercise portion of your graduate life, now is a good time. What are some of the specific steps you can take to incorporate exercise into your life?
3. Talk to some of your peers and note how they incorporate the well-being of body and mind into their lives.
4. What are some ways you can help the graduate student community in your department bond?

Notes

5

Organize

Now Is the Time

Whatever you do, do it with excellence.

Maybe you've just walked through your undergraduate graduation ceremony (and then a beach holiday). Maybe you've decided to come to graduate school after disillusionment or boredom set in as part of your job (that was my case, a long time ago—the boredom part!). Or maybe you've been told that getting a graduate degree is definitely a path toward upward career mobility! Whatever the case, you have decided to take the plunge. You want an advanced degree. You've picked your university. You have e-mailed your advisor and possibly even met them once or twice. You've checked out the local accommodations. You have made sure that your graduate research assistantship is on par with other universities. Now you are here.

This is where you will spend the next 3 to 5 years (or more) of your life. No, I don't mean this to sound like a prison sentence! Chances are there are several other fellow sojourners also arriving the same week with goals similar to yours. Trust me when I say this: your graduate school days are some of the best in your life. I am sure that you will say the same thing about graduate school after you have received your degree and after you have spent several years in the workforce. For the most part, being in graduate school with or without a research assistantship is a free and footloose situation with very few obligations. The money may not be great, but graduate school life should be fun! For that matter, even if you are not being paid to go to school, this should be a fun experience. There are not many places in the world where you can go to school to get an advanced degree, discover things, and possibly even get paid for doing it!

You may be an organized person by nature—in that case, read this chapter anyway—or you may have gotten away with sloppy organizational skills during your undergrad days. Graduate school, as I often tell my students, is not hard. Half the battle is staying organized. I seem to be giving this lecture quite often, because when I confront students about the lack of productivity, the first thing that I hear from my junior students is that they have a lot going on in their lives. They always state that they don't have enough time. They come up with a huge list of things that prevent them from being productive in research. Rather than telling them, "Tough luck—you simply have to cope with it," I have started to get a little more philosophical, or sophisticated, in my response. After all, I have a lot more gray hair than I did when I started my career!

I am often reminded of the late Larry Burkett, a favorite financial planner of mine. He used to relate this episode when he started counseling people regarding budgets. He would ask for income and expense numbers from his clients, and they would show him the budgets that proved their point: they are in debt because they borrowed money and there was no other way to make ends meet. From the couple or individual who made 20K a year to the individual who made 50K a year to the couple who made 250K a year—they all attempted to prove to the financial advisor why they are in debt. They simply could not make ends meet since they were not making enough money! The financial planner quickly came to the conclusion that it is not how much they made that fully determines the debt scenario. It's how they spent it! You have to give up a few things to stay on course—to stay out of debt.

Now I draw that analogy from financial management, but the same applies to time and project management as well. We all have 24 hours in a day (although some make it look like they have more!), and whether you are a student or a professor, chair or dean, director or president, you have to make sure that things get done right and on time. Otherwise, things go awry—not only for you but also for others who depend on you! I boldly tell my students that I manage at least 20 things more than they do (and my boss manages 20 things more than I do, etc.), and I have to be accountable to my boss. I also tell them that I couldn't give an excuse to my boss about the lack of time. They seem to get the message that poor time management is no excuse for sloppy work and for providing perennial excuses. They usually get the point that time management is essential!

Journal Entry

I will purpose in my heart to manage my time better—every day! What methods are you putting in place now?

Food for thought

1. List at least five challenges that prevent you from staying organized.
2. List your strengths and weaknesses related to organizational issues.
3. Think of ways you can mitigate weak areas and build strengths.
4. Observe peers and mentors around the area. What can you learn about their organizational capabilities?
5. Set up an appointment with an "organized individual" to learn more about their methods.

Manage Your Time

You must give account of your time.

Time management. This very phrase conjures images of electronic-gadget-driven students and professionals with robotic lifestyles complete with pocket protectors peeking through shirt pockets. It doesn't have to! Time management is a critical skill to be learned and mastered during graduate school. Granted that you may never conquer time management, but you definitely still need to work at it. If you have a time management system in place, this is the time to refine it as you anticipate your career after graduate school. If you do not have one in place, you had better develop one—now!

It all seems simple at the onset of your graduate career—a few classes, one or two projects, and some papers and books to read! Very soon, it multiplies to writing lots of project papers, working on more challenging projects, teaching some classes, being a teaching assistant, and managing your dissertation. And don't forget about a personal life! As your time in graduate school increases, your responsibility increases as well. It should—otherwise your advisor is not doing their job. They should be providing multiple tasks and increasing your responsibility levels to prepare you for *your* career. The list of responsibilities keeps increasing. I've seen this scenario over and over again. Papers are strewn everywhere. You're late for classes sometimes (and in some cases always). Your projects are getting sloppier, and the cracks are now visible. What's wrong? Time management!

I am not even going to attempt to tell you how exactly to manage your time, but I will discuss some common drains and how to better stay on top of things. Even the best time managers will state that they could do better. The reason that it is important to learn how to manage time now is to prepare for the high expectations once you land that 100K per year job (just kidding!) after you graduate. There will be more things to do than you possibly could handle without good time management— and more will come. Graduate school, in my opinion, is the best to place to flex your time management skills, learn how to prioritize, and finish work and projects on time. Put a system in place, and then you can refine it as you go along.

I have often asked my students to list the challenges of managing time effectively. I also asked them about significant time drains that hamper proper time and task management. Given that we live in a technology-driven world, it is no surprise that I hear that computers are the worst time drain. I suppose it used to be television and probably still is for many students. I don't want to sound like a parent admonishing a child for the misuse of computers and television, but it is a fair warning that if you give your energy to something that does not promote productivity, it will drain your time. E-mail and chat programs on the computer come up as another source of time drain. The list keeps growing with time: Facebook, Twitter, etc.

My recommendation is simply to turn all of these gadgets off until a set time each day. Then you can open up these tools—for example, your e-mail—and take care of business. I know of a successful researcher who does not open up his e-mail until after lunch. He checks and answers his e-mail for about 1 hour and then shuts the e-mail component of his day completely off. In this way, a pattern is established and people know when they can expect an answer. Once a day! I agree that it may not work for all professionals, but at least it provides a great tip on how to allocate and manage time for e-mail. The most important thing is to identify time drains and come up with plans to negate them. No two students' time drains are alike. Therefore, you need to give this some serious thought!

Another common and less identifiable time drain for a graduate student is office talk. Often, students share offices with one another. If you don't watch out, your office could become a hub for gossip, frenzied activity, marketplace discussions, or hanging out. Learn how to politely steer the discussion to a common place.

> **Survival tip:** Stay away from anything and everything that smells like departmental affairs or gossip. Discussions about who the faculty are dating or having lunch with and how the department secretary behaves at meetings are counterproductive and quickly become a time drain. Focus on your end goal and let the trivial work go down the drain by itself.

Take-home message

 The first task in any time management exercise is to identify what the tasks are for a graduate student and then to identify the possible time drains so you can do something about them. Time management is not just about putting a schedule into a gadget or on a calendar. It is a deliberate, daily to weekly activity of how to effectively manage your time.

So how does one manage time and activities? I have had a range of responses from students and colleagues alike over the years, and rightfully so, since no "one size fits all" strategy will work. I have asked several of my colleagues how they manage their time and their tasks, and I have received various responses as well. Some (me included) hate for their life to be run by a gadget that has to be in their pocket at all times. Others (me included) are vehemently opposed to having a lightweight laptop run their lives because they have to carry it everywhere. Some don't manage things well at all and seem to forget meetings or always arrive late for appointments (I don't like that at all!). Others still carry a small calendar/diary in their pockets (hmm!). I had one student who indicated that she does not manage time at all, and she doesn't believe in prioritizing or itemizing. Needless to say, she had a difficult time coping with the rigors of graduate school and was a constant source of frustration and irritation to her advisor, her professors, her peers, and herself. She took nearly 8 years to finish a Ph.D., and an unhappy Ph.D. at that! On the other end of the spectrum, I have had students who purposefully allocate time (to the hour) for exercise, research, classes, etc. I admire those superorganized individuals.

One of the most interesting approaches that some students and faculty employ is what I call a staggered approach to prioritizing. They put together a prioritizing method by itemizing every morning (while in the shower for me). Some also have a weekly list of activities on paper somewhere (from sticky notes to organized calendars). I am sure that management of activities and time evolves with responsibilities, as it should. Otherwise, life would get either chaotic or downright boring. Again, this varies from one individual to another. If you are forgetful by nature, you had better have a strict system in place so that you do not forget to show up for key appointments! It is annoying when students show up late—repeatedly—for appointments. This is definitely not a good way to demonstrate motivation.

Learn how to prioritize and work on a system that helps you. This is like that old nag of a word called budgeting. We all hate it but know we have to—unless we have an infinite amount of money. Time is the same way. If you don't pay attention to it in graduate school, it really runs away from you—and remorse and regrets don't help at all!

How do I manage time? This is one of those questions I get asked a lot. Well, when I was a graduate student, I did the itemized list approach and meticulously prioritized by various symbols and colors. Then I used sticky notes in strategically placed spots; for example, the sticky note on the top right corner of my computer was the most important, and the one on the left desk drawer was the least. I soon got tired of that, and this later evolved to what I thought was a brilliant idea of having a busy board: a white board with various color markings of activities. That too became tiring.

Now I think I have arrived at the greatest idea since sliced bread: putting it all in an online calendar with appropriate notes. The people in my life who really need to know my whereabouts also have access to this online calendar as well. More recently, I have succumbed to a "smart phone." I hope to see how this evolves with time. With added responsibilities comes a need to invent something that works for you. Be creative. Rather than thinking of managing time and projects as a chore, have fun with it.

Having said all of this, all I want you to think about is a system, any system, to start out with—prioritize, make a list, and stick to it. Modify it until you arrive at something that works for that phase of your life or your career needs. Iterate as necessary. I am sure that my time management system will evolve very soon—maybe to one without any appointments. I can dream, can't I?

At some point, you may become so rich and famous that someone will manage all of your activities! Won't that be nice!

Journal Entry

List three challenges/time drains. What are some specific plans for minimizing them?

Food for thought

1. *For the instructor*—Ask how students manage their time. List them on the board and talk about pros and cons.
2. Conduct a SWOT analysis on time management and list top three factors that could easily provide additional time for building areas of strength and minimizing weak spots.
3. In your opinion, who is the best manager of time? Pick one student and one faculty/staff member. What do you think is their strategy?
4. Talk to good time managers and see if their input will help you to adapt their strategies or blend in with yours.

Beat the Stress

Press on toward the goal in front of you.

There is definitely a link between poor time management and stress. When time drains are the norm in your day, it very quickly leads to stress factors in your life. As a graduate student, it is easy to procrastinate on projects—at least I did. Even if there are only a few weeks left for the finished product, we all tend to overestimate our ability. Therefore, we start late and still expect to finish well. Since most professors also know this path from student to professorship, they are good at grading sloppy versus thoughtful papers. So beware when you try to start late because the quality and rigor will suffer.

When I ask my students questions regarding stress factors in their lives, I almost always get a vast majority who mention poor time management. It is not that we don't know that putting things off will make us miserable. We still do it though. To make matters worse, poor time management on your part will have a cascade effect on others as well since in your future job (or even in graduate school) you will be a vital link in projects. When one link breaks down, everyone suffers! Nowhere is this more evident than in a team environment. When someone drops the ball on a project, others around them will have to make up because of the cascade effect!

As a graduate student, it does not matter if you are in a semester or a quarterly or whatever system. Projects, exams, and deadlines all seem to appear about the same time. Don't they? What I am going to suggest is downright simple, but it provides some guidance anyway. Mark all deadlines on your calendar and make sure that it stares at you at all times. Get one of those calendars that have multiple months on one page so you can see your deadlines for the entire semester at a glance. An online calendar on a computer will work just as well. To get yourself really going, mark the percentage of points that each one of those tasks carries for that assignment/project. For example, your final project in a certain class may carry a 50% weighting. If that does not get you going early, then you are in trouble! It doesn't matter what I say at that point, you will be in serious trouble if you started that project a few days before the deadline.

Research deadlines are largely self-imposed. They should be. If you expect your advisor to get *your* act together, then you shouldn't be in graduate school! You must make progress each week so it is fulfilling to you and so it also satisfies your advisor. We all like to experience success, and you have to put a system in place that will be meaningful for you. Remember that you have to take small steps first before you can run. This is important when you are doing research, because some, if not all, of the topical area of interest may be new to you. It takes time to read, assimilate, and learn some new tools and analysis skills before you can really produce. Take heart, your advisor knows this as well. If they expect you to produce from day one, then this might not be the advisor for you.

I've tried to make this book as practical as I can without making it sound like a recipe book for success. I don't think a recipe book for success even exists. However, I have been asked how a student should progress in research when given a 4 year timeframe that includes taking various classes and exams. Most students do not seem to have any problem in maneuvering coursework since there are

Table 1. Time Management Guideline

Year	Activity
1 (starting in fall)	Take classes.
	Build tools (take tool-building classes).
	Fill gaps and weakness in background if necessary.
	Read papers.
	Build a weblog and a system.
1 (summer)	Take a programming class.
	Ramp up research.
	Take preliminary exam, if applicable.
2 (starting fall)	Take classes.
	Refine tools.
	Write first paper on broader research.
	Take classes.
	Attend conference.
	Accelerate paper reading.
	Refine and build blog.
	Take qualifying exam if applicable.
3	Take light load of classes.
	Take communication courses.
	Attend annual conference of choice.
	Write papers 2 and 3.
	Continue blogging and refine your resource center.
	Mentor younger students.
4	Writes papers 4 and 5.
	Take light load of classes.
	Give seminar to larger group.
	Attend conference and network.
	Defend dissertation in April–May.

specific deadlines and exams. But when it comes to research, there seems to be very little guidance on how to progress.

Let's assume that the student has a plan to finish their Ph.D. in 4 years in a three-semester-per-year system. Then the basic elements are building tools, taking classes, passing exams such as preliminary and qualifying exams, proposing a research topic, writing papers, and defending research.

For you to effectively complete this schedule, you need to draw a map that includes a first line with an academic schedule of classes and exams and associated deadlines. On the second line, draw a plan for your research. On the third line, draw up a plan for major personal events in your life. Some students plan weddings during their graduate school, while others even take time out for maternity leave. Whatever your personal calendar is—map it out and have a plan of action. Table 1 provides a sample guideline (merely a sample!)

Journal Entry

 What are some current stress factors and some potential future ones? Write down your plans to conquer the stress factors.

Don't Get Weary

Don't be weary of doing well.

Let's get this straight. Going through graduate school and planning and positioning for your career is not easy. It requires careful, diligent, and purposeful planning. I hope you take away this message from this book.

Have you ever run a marathon? Well, I haven't, but I have seen some do it. I know some of my colleagues and peers who run marathons on a regular basis. They talk about the highs and lows of running 26 miles. They talk about getting in the zone and thinking about the finish line, but they also talk about an all-important thing: the fatigue that sets in. But guess what, they have practiced long and hard, and they know that they have to get the correct mechanics of running to get to the finish line. Graduate school sometimes seems like a marathon. I know for certain that there are lots of highs and lows. Some of you reading this book have just completed your undergraduate degree and are embarking on a graduate career with the intent that you want to finish your graduate work once and for all. Others of you have come back to reenter the education pipeline after some dabbling in the workforce. Some of you are in graduate school full-time; others are part-time students with a part-time job. Some of you have to foot the bill to go to graduate school, while working among others who are given a scholarship or a research assistantship of some kind. There are whole ranges of other scenarios.

Let me see if I can paint this picture a little bit better. Here's a typical scenario: Graduate school seems to be going well. Classes are humming along. You are picking up a lot of valuable tools, and research is progressing fine as well. You have your markers and milestones identified, and they are so tall that you can see them for miles and years out in front of you. You are not walking or running toward your milestones, but you are driving toward them—at Porsche speeds.

Then you hit a roadblock, a massive one. Anything can become a roadblock in graduate school. It could be physical or emotional issues, problems with your colleagues or advisor, or simply the graduate school blues. Ever been there? Whatever the scenario, I have rarely come across the individual that has breezed through graduate school without missing a step. If you are one of those students, you should be writing this book!

Even after all these years, I still remember distinctly walking into my office a long time ago when I was a graduate student and venting to my officemate that I was tired of the process, and I had no idea whether I was ever going to finish. Guess what? We all get in desert places like that at times, and it happens to most, if not all, graduate students. If you are one of those in this situation, then this section is for you.

There are some traps that you need to be aware of when you are in this desert place. If you surround yourself with people and influences that encourage a downward spiral, then it is going to be all the more difficult to get out of the rut. It is quite alright to be philosophical and vent for a while, but never let it get out of hand. Think marathon! You have to put one foot after another, slowly, deliberately, but surely until you get back to that zone. Think in short segments that will get you to the finish. Learn how to surround yourself with colleagues who will provide water to the desert—not throw in more rocks and stones or ask you to read downward spiraling philosophical junk theories. You are not going to get water there—merely disillusionment.

At the risk of sounding too harsh, I'll relate this story. A while ago, a student in a midsized department got into one of those ruts. Instead of thinking about mile markers and finishing her degree, she began listening to folks who began to philosophize about her situation. Very soon, she was ranting and raving about the meaninglessness of all of the exams and the research. Beware who you listen to and what you read when you hit these bad spots. Think positively!

Just a short while ago, it seemed that you sped by mile markers in your Porsche, but you now seem to be limping. The milestones are not visible anymore and everything is hazy. It seems like you might be walking in circles. What should one do in this scenario? The first thing to realize is that you have to get out of this desert place, or, for lack of a better phrase, snap out of it!

Enter milestones or markers. You have to think about the various graduate school markers that you have put in place, and you have to steadfastly move toward those markers. That's why marking those milestones, success stories, and strategies that you have put in place in your journal could be useful. When you are in a desert place, you can go back and revisit some of these strong points in that journal.

My advice for students who hit these desert spots is to step back and look at the bigger picture. It is important to know what the next marker is and how far away or how close you are to it. If it is a research problem and if you are stuck in a desert place (a difficult problem), you have to think of other ways of solving it. Read the literature and try different things. A breakthrough will eventually come.

Have you ever been in a situation where a piece of computer program that you wrote was working, and it now screams with multiple errors every time you run it? Oh, yes! I have been there myself. Frustrating as it is, you must realize that this too will pass, and you have to simply retrace your steps one at a time until you solve the issue so you can move forward. Whatever the situation, you need to focus on the goals ahead of you!

Personal issues can also set you back sometimes. A parent who is diagnosed with a life-threatening disease or a personal health failing. Stuff happens! While these things are serious in nature, you have to simply take a step back and reevaluate priorities before you move forward.

The main thing to remember with setbacks and desert places is to figure out the markers and slowly get to the next one and the next until you reach your destination. Draw a plan and set mile markers for every 3 to 6 month period of your graduate career. Assume that you have 4 years from start to finish. Separate this into academic, research, and personal markers. Once you have mile markers in place, you can navigate life a bit easier. Again, knowing where you are going helps in the preparation process.

Journal Entry

What constitutes setbacks for you? How will you handle them?

Food for thought

1. List some of the setbacks that you have or you might encounter during your graduate career.
2. Who might you turn to if you find yourself in a desert place and what are your expectations from this person?
3. If you have not hit these grad school blues, list some ideas about how you might tackle them if you do.
4. If you have a serious health failing and 2 years of school remain, what you will you do?
5. What role do you expect your mentor/advisor to play in this process?

Notes

6

Writing

Ditch the Dissertation

Count the costs.

Whenever I think about the tome called the dissertation, it conjures up images of about 300 plus pages of writing, toiling, and slaving away at the changes—and then more changes! Sometimes I feel that students take pleasure in stating that their dissertation has the most number of pages in the history of the department. First, you write. Then the advisor makes changes. You continue to iterate until your advisor is reasonably satisfied (are they really?). Then your committee reads it, and more changes are made. Finally, the crowning moment of your defense arrives. At your defense, you stand in front of a lot of people, talk about your research, and field questions. Later, the university coronates you with a degree. OK, not so dramatically, but that is the sequence of events. Very few external peers have read your dissertation at this point, and, after you have defended your work, you have the ominous task of now slicing this 300-page document once again into smaller-sized chunks and turning them into journal papers. I bet a lot of you are nodding your heads at these statements and saying—been there and done that! Have you ever asked yourself this question: Why?

Why should you go through the agony of writing a 300 page document just to sit down and chop it up into two or three chunks for journal papers? After all, it's only the journal papers that count, right? Nobody ever reads your dissertation. I am still waiting for someone, I mean anyone, to ask me for a copy of my dissertation. I have five extra copies of mine on my bookshelf! This grueling exercise of writing a dissertation must be some sort of tradition or at least a requirement!

So let's ditch the formal dissertation, shall we?

I often ask students to do something so radically different that I have to sometimes explain several times, slowly, almost as if I am trying to sell a pyramid scheme of buying laundry detergent. I am surely not the first one to come up with this method of writing a dissertation, but it sure makes life a lot easier, for both you and the advisor!

Here is how it works. As you are doing your research, you come up with one overarching theme and then compartmentalize your research into two or three separate sections. Let's assume that you can slice your dissertation research into two sections. Some call it part 1 and part 2, but I prefer to simply compartmentalize it. A year before the final defense, you can slowly start submitting the two sections one after another to the same peer-reviewed journal. After the first one has been submitted and is in review, you write the second paper. By the time the second one comes around, you are getting the reviews back from the first. This is a cyclical procedure, and it is also a rewarding process since these comments are from experts in your area of research.

As much as committees mean well, not every member is necessarily an expert in your area of research. Therefore, getting reviews back from the journal experts carries a lot more weight. Hopefully, before the defense date, you will have two to three papers accepted and in press. All that matters is that you were able to compartmentalize your research into sections and have all papers accepted for publication. Now here is the fun part. You can completely ditch the arduous dissertation writing process. Remember that these papers have been carefully examined for scientific content, clarity, and technical robustness by the experts in your area. Your advisor has also worked with you on content, style, and grammar. It looks even better if some of your committee members are coauthors.

Here is how the assembly works. Each paper usually has a title, an abstract, data, methodology, results, and conclusion, with references, figures, and tables. And here is the best part. Each one of these journal papers that you have written and submitted has been spell-checked, proofread, grammar-checked, and attested by peers. It is certified by a journal as being original and of high quality. I make it sound like it is meat that has been inspected, cleared, and certified by some Food and Drug Administration officials. Bear with me.

Nearly 90%, if not all, of the writing is practically done for your dissertation! You don't believe me, do you? Let me explain.

You simply have to take the two or three peer-reviewed papers and assemble them in a form that the university regulations stipluate. I usually ask my students to simply take the documents and begin to rearrange the sections. First, come up with one grand title for the dissertation. Seam all three abstracts together into one general abstract. Edit lightly for proper flow of sentences and sign your name underneath it. Alphabetize the references with a few clicks of your favorite word processing program and stick it at the very end of the document. Write a very simple two page overarching introduction by providing an overview of the three "subsections" (the three papers) to follow. Now simply place all three papers back to back. In the very last section, write a page about some of the future work that could be done. For the overachievers and show-offs who write five or more papers, you can seam all five papers together, or you can add those two extra papers as an appendix to your dissertation.

You are all done, and it is completely legitimate. You have satisfied several things at the same time. Your resume went up several notches. Your advisor can now clearly point to your papers as a success metric. The department and the university have a top-notch dissertation, and you did not have to go through the procedure twice—once for writing a dissertation and the other for turning them into journal papers.

Not only do I insist that my own students use this method, but I advise other students to use it as well. As an advisor, I want to be spared the task of reading a 300 page dissertation and then having to go through the same process again to turn it into journal papers, waiting agonizingly to see how the reviewers respond.

When I present this in the class or during discussions, most of the senior students who have already embarked on the dissertation journey groan and mutter indicating that they wish they had known this earlier. Now you know!

This approach is advantageous for several reasons. It cuts down on a lot of wasted time writing unnecessary and irrelevant things in the dissertation. Though this is a crowning achievement of a student's career, it is still painful to read a long, drawn-out dissertation. I am and continue to be on numerous student committees, and it is indeed painful to hear the student say that they will write their papers **after** their dissertation.

By writing everything in a journal format, it provides conciseness and a focus for the student and the advisor. Since most of you will be writing lots of journal papers in your future, this is great practice. Given that the papers are reviewed by experts in that particular area of interest and the fact that your committee members in your department are all not experts in your dissertation area, this approach for "ditching the dissertation" is important.

Finally, on that all-important day when you stand up to defend your work and present your research in front of your peers and your committee, the first slide after your title should list the titles of all the papers you have written and, in bold, the words "in press." At that point, I don't think anyone is going to argue with you much in your defense, let alone ask you to make changes. The committee then fully knows that you have already gone through the process in the peer-reviewed world and you have succeeded. If anything, your committee members (CMs) will be jealous!

Finally, I have seen too many students not use their thesis or dissertation to write papers. This is a colossal waste of time, in my opinion. However, the effort of reassembling a dissertation into a paper can be daunting. The student oftentimes moves on to another job, and the dissertation research is forgotten. This is another reason why you should write journal papers and not a traditional dissertation.

Learn how to compartmentalize your research. Write at least three papers before graduation and seam them together. Trust me when I say this: even you will not read your dissertation after you graduate. And it is safe to say that, apart from the committee members, no one else will. Send me an e-mail, and I will send you a copy of my dissertation. I have five copies, remember?

Some thoughts on compartmentalizing research

Over the years, I've had to politely (!) disagree with the statement that "Compartmentalizing research is diluting the rigor and completeness of research." As nice as that sounds, students lose focus with long, drawn-out procedures. Traditionalists often state that a student should write a complete dissertation.

After they defend their work, they should take some time out to rework their dissertation into one or two journal papers. I simply don't buy into that theory, and it hardly makes sense, especially at the pace research is moving along in most areas. I feel that writing papers and then assembling them later into a dissertation makes much more sense. Traditionalists have to get used to the idea that the way we do business has to change as generations change. It has nothing to do with sacrificing quality of research or rigor. Newer generations expect, and oftentimes demand, definable end points, success metrics, and usefulness of research. As advisors and mentors, we had better get used to it!

Journal Entry

I will purpose in my heart to write three or more papers before I graduate.

Food for thought

1. Would you really like to write a formal dissertation as opposed to seaming papers together?
2. What are the disadvantages of writing peer-reviewed publications before you graduate?

Write a Paper

Write a paper. Know what counts.

Writing papers for peer-reviewed journals is a long and arduous process for most, but in my opinion, it is most rewarding. It provides focus and a definite end point. The first end point is to send the paper out for review after the writing process is fully completed. This is a tall order by itself. Closure is achieved when you see a paper with your name on it in a journal. As a seasoned veteran put it, publishing is just about synthesis. When you publish well (and frequently), it means that you know how to manage time, projects, stress, life, research, your advisor, and your environment!

I hope that you are still smiling from my section on ditching the dissertation where I wrote that no one ever reads your dissertation except your committee. Therefore, writing high-quality papers is important. They must be peer reviewed or it doesn't count. Many students and researchers often make the mistake that only quality matters and quantity of papers is irrelevant. I disagree strongly! In an age where research is progressing at a staggering pace, it is not enough that we produce one or two papers over a 3 to 4 year period. For most areas of research, rapid dissemination of results is not only the norm but is becoming a requirement. I have not done a survey, but I would dare say that promotions and tenure are given to candidates who have demonstrated productivity through numerous peer-reviewed papers rather than one or two quality papers over a long period of time.

You cannot get away from this: wherever you are, there ARE bean counters. Your boss and your system have metrics in place, especially for young investigators and researchers. The number of papers matter and, if you don't count them yourself, then someone will count them for you. I have been on enough committees and panel reviews to know that productivity in terms of not only quality but also quantity is also discussed.

A while ago, I interviewed for a position at a prestigious national agency, and one of the members of the selection committee opened up a folder and told me about *my* publication record—quality/quantity and how I stacked up against national averages, regional averages, and scientists who were in a similar position to mine. Even I hadn't done such an analysis on my own publication record! As I mentioned before, bean counters are everywhere!

Having said that, I know the question you are dying to ask, especially if you have already graduated and are in a job. How many papers should you publish each year? I cannot answer that because it depends on your discipline, the norms for your department or group, and the time it takes to actually see a paper in print after it has been accepted. The important thing is to write papers at a regular frequency. This indicates that you know how to carry things to completion. I suggest that you do the following. First, talk to your supervisor or manager and find out what the expectations are. Then talk to peers and colleagues in and around your disciplinary area. Better still, check out similar scientists in your area of expertise at other places and see what a successful metric might be. Very soon, you will get a good handle on how many papers to publish per year. The simplest thing is this: Try to publish one more paper than a successful member in your area of expertise, organization, or department who is at the same level of experience. If they are successful with N papers, $N + 1$ or $N + 2$ should be better. Correct?

If you are a student, I suggest that you write two to three papers in broad areas regarding your research and then write two or three more in your actual dissertation area (if you are a Ph.D. candidate).

I have talked about compartmentalizing research so that you can write papers. If you cannot do that, then you become easily frustrated since you never seem to get your papers out the door, even for the review process. One of the frustrating types of conversation that I have is with students who have a lot of potential but never know how to write and finish a paper. Year after year when I see them in the hallway, they mention how many new ideas they have, and they have mapped out everything in cerebral space. Guess what, no one can get into your head and read about all those brilliant ideas. I tell my students in class that unless you have a peer-reviewed printed or e-paper with your name on it, none of your brilliant ideas really count. But the first task is to get the paper out the door. That should be a mile marker!

Before you write a paper, you must know which journal is most appropriate for your submission. This is definitely a place where you need guidance from your advisor or successfully published peers. Journals vary in length, quality, and rating. The most important thing is for you to concentrate on writing a high-quality paper for a reputable journal in your field of expertise. Often young researchers worry about the impact factors of flashy journals rather than focusing on high-quality work. The idea is not to gain recognition with flashy science but with a solid and thorough piece of scientific work.

Once you have decided on the journal, your task is to read through the journal Web pages on page limitations, number of figures, and requirements. At this point, you are ready to write a paper. It is difficult to generalize on how to write a paper, but I will provide some basic pointers.

You should consult the specific journal for your chosen field for further information, but most papers have the following format:

- A concise ***abstract*** that discusses the major focus areas, results, and implications. Many people only read the abstract, so you need to make it relevant.

- An *introduction* that provides a relevant literature survey and discusses how this piece of work is important.
- *Data and tools* sections that outline how they were used to conduct the research.
- *A methods* section that provides the reader with enough information to replicate the results if needed.
- *Results and discussions* with figures. This consitutes the bulk of the paper.
- *Summary and conclusions* that put the results, discussion, and analysis in context and presents some guidelines for future work if possible.
- Of course, an *acknowledgment* section and *references* are also needed.

As a student, sometimes it is a daunting task to write that first paper because the tendency is to want to include too much information. Most students (the good ones at least) come to a meeting armed with at least 50 figures to discuss the elements of a paper. I expect this because I know they are eager to show **all** their results. As an advisor/mentor, it is my responsibility to walk them through the process of compartmentalizing their research and making sure that they understand the important steps of writing a paper. I warn the students in my class that, even though one could seemingly put 50 figures in a paper, no one, especially these days, has the attention span to read through all that information. The trick to writing any paper is to decide upon the appropriate theme and the figures necessary to illustrate a good storyline and thread. Often the best writers have a story that they tell elegantly regardless of the necessary scientific jargon.

It is a good idea to put page numbers and line numbers as this helps the reviewer. Also, remember to check spelling, grammar, and references (both for in-text citations and the reference list) before submitting the paper. It is often customary that the paper is double-spaced and in 12 point font with a header that carries a short title.

Following the title page is the abstract, introduction, data, methods, results/discussion, summary/conclusion, and acknowledgement (who funded the research and other acknowledgements) sections. These are followed by references, list of tables, list of figures, and the actual tables and figures. Most journals like to have the figures separately at the end of the text rather than embedded in the text. Again, you should read the paper submission guidelines for the particular journal before submitting your paper. Some have page limits as well as limits for the number of figures and the number of words in the abstract. Once you've been through this process a few times, you won't need to read this section again. But we all have to start somewhere.

OK, let me break at this point to briefly mention how the review process works. Keep in mind that the outcomes will always enlighten the writing process. Once you have written a paper and submitted it (usually all online these days), the review process begins. However, when you submit a paper, make sure that you include a cover letter mentioning the title, the authors, and your brief opinion of how this paper is appropriate for the journal. Some journals require you to specify at least five potential reviewers with e-mails as well.

Once a paper is submitted, the editor of that particular journal along with an associate editor (optional) will take over the process. A good editor often makes an informed decision about who should review the paper. Some journals have three reviewers, while others have only two. Once the reviewer agrees to review the paper, they may have from about 10 days to nearly 1 month (sometimes even more). The evaluation criteria are not as stringent as a proposal. The major objective for a reviewer is to determine if the paper is scientifically and technically worthy of publication. Obviously, you should have done your homework to make sure that you have submitted a paper to a journal in its area of readership and the topical area of interest. There are always some other review criteria such as, "Is the paper written clearly" and "Are the figures of good quality." None of them are really showstoppers unless of course you have done a terrible job of proofreading or used hard to read graphics. If the paper is technically solid, then it should be accepted for publication.

There are several choices for a reviewer: reject the paper outright, accept with major revisions, or accept with minor revisions. Note that while the reviewers can recommend a decision, the final decision rests with the editor. After all the reviewers return their evaluations, the editor then reads the reviews to see if they are consistent. It is easy for the editor to make a decision when all three reviewers agree on whether to reject or accept the paper. It becomes difficult when the reviews are mixed.

At this point, the editor can send you the decision either to simply reject it or to ask you to resubmit a revised version taking into account the comments from the reviewers. (Note that this process is anonymous. You have no idea who the reviewers are, but you do know who the editor is.)

Scenario 1: Paper rejected

What should you do when a paper is rejected? Like a proposal, you should regroup, read the reviews carefully, rework the manuscript according to the reviewer comments, fix flaws in the paper, and resubmit to the journal. Talk to the editor and discuss your options. Sometimes editors may suggest that your paper, while technically good, is appropriate for another journal. There is also nothing wrong in resubmitting the paper to another journal only AFTER you have made the necessary changes. It is a poor practice to simply take a rejected paper and resubmit it to another journal without adequately addressing the issues. Tempting, but poor practice! Remember that as good or bad as the reviews may be, you should give all comments proper attention. The reviewers have taken the time to read through the manuscript, and it is safe to assume that the reviewers are experts in the area of research.

Scenario 2: Accept without any revisions

If your paper is accepted without any revisions, then it is akin to winning the lottery. In all my years of writing papers, I never have had a paper with zero revisions. In that case, the paper is usually sent from the journal to the production office for typesetting and publishing.

Scenario 3: Accept with minor revisions

Another possibility is that the editor sends back the reviews and tells you that the journal is willing to accept the paper if you take care of some minor issues raised by the reviewers. This is probably the easiest and most joyful of all experiences. You simply address each reviewers comment in a point-by-point fashion. When responding to reviewer criticism, be polite and do not use critical or harsh language. Avoid phrases such as "The reviewer has no clue about this concept." As tempted as you may be to be harsh, refrain from making such arrogant statements. It will only alienate you from others in the research field. Be concise and to the point in your responses. After you have responded to each criticism or query, you should write a short cover letter stating very briefly what the issues were and how you have addressed them. Once your responses and your revised manuscript go back to the editor, it is usually accepted, and a letter is sent back to you indicating that the paper has been accepted. The production office for that journal will contact you for further information and guidance.

Scenario 4: Major revisions

The other possible scenario is that the editor calls for major revisions. This means more work often in terms of redoing some part of the analysis, analyzing more data, or providing more input. You are given a reasonable amount of time to do this. I usually consider this a learning experience. The reviewers usually provide good feedback. You should carefully consider their viewpoints and strive to address those issues. This is what turns your good paper into a better one. You now have experts providing valuable insight that you should address. Never shortchange yourself by ignoring good comments. In the past, I have taken into account reviewer comments by analyzing an extra year's worth of data to answer a piece of the puzzle. It takes a while, but the results are worth it.

For a major revision scenario, the paper is usually sent back to the reviewers (all or a subset of them), and the paper is reviewed again. The process is repeated until the reviewers and the editor are fully satisfied. At that point, you will get a letter stating that the paper is accepted and to await instructions from the production office.

I have written about these steps in some detail since many of the students that I have encountered do not have a full understanding of what is involved in the paper writing, submission, review, and production processes. Keep in mind that not all of these steps will apply to your disciplinary area. It may be slightly or totally different. I cannot possibly cover all disciplines and the nuances of the submission and production process.

Next is the production process, which is often painless. Once the paper is accepted, the journal's production office often contacts you to request the graphics in the correct format and resolution and a text file that they will typeset to the journal's specification. After this, you can pretty much relax and follow the process online regarding the status of your paper. The day that your paper appears online, you will usually get an e-mail that your paper has been published. It is a wonderful feeling to see your name in print. As happy as you might be, you still have responsibilities. Your job is still not over!

Most students look at me strangely when I ask what their responsibilities are once the paper has appeared in print. Remember that hundreds of journal papers in various journals all across the globe come out each month—even in your area of expertise. While you may think that the whole world has been waiting with eager anticipation for your paper, it is rarely the case. Other than some of your friends and colleagues, the editor, and the three or so reviewers, no one even knows about the paper. If you think that all researchers look for papers in their field on a daily basis so they can find your paper, then you are in for some disappointment. Often good papers go unnoticed!

So you have some public relations work to do. If your research is funded, your first responsibility is to e-mail your program manager (PM), the one who funded your research, or send them a hard copy of your paper if they prefer. Include a short statement of the highlights of the paper. Next, you need to send it to your colleagues in your field around the world who will read and reference your work. Then you need to inform the chain of command wherever you are (and those bean counters) that you have a new paper that has come out. You are still not done. If you have not been presenting your paper in conferences before publication (bad practice!), then you need to at least make sure that you do so now.

I cannot say this enough: Learn to write journal papers as you are doing your research. Think of your dissertation as one seamless idea that can be compartmentalized into smaller pieces. Experience success while in graduate school by writing papers!

Journal Entry

I will submit my papers on the following dates: _____

My papers were published on the following dates: _____

Food for thought

1. What are some stumbling blocks for you to write your first paper?
2. Do you know how to compartmentalize research? If not, who would you talk to?
3. List three journals that you could possibly submit your paper to and who the potential reviewers might be.
4. How does one go about picking reviewers for a paper?

Notes

7

Proposals

Why Write a Proposal

Since this book is about all things practical, here is another great idea. Learn how to write a proposal while you are in graduate school.

Why should you learn how to write a proposal? There are a myriad of reasons. The obvious one is the fact that you may have to write proposals for a living after you get your advanced degree. Proposals are nothing but a sales pitch. You must have a good idea and package it well. What better place to learn that than in graduate school? When I ask students for a show of hands as to how many know how to write a competitive proposal and prepare a budget to go along with it, I get some pretty strange looks. Yet these are some of the students who will be forced to write proposals very soon after they graduate.

Some of my colleagues are going to mutter under their breath that I am going to give away secrets to good proposal writing. Yes, that's exactly what I intend to do in this section! Before I get into the nuts and bolts of proposal writing, here are some pragmatic considerations. If you are a Ph.D. student, then you have to write a research proposal and provide this to your advisor and committee members before you embark on your dissertation project. Why not pretend that you are going to submit this Ph.D. proposal to a federal agency that awards competitive grants? Also, pretend that your committee members are your reviewers and panel members. If you then pattern your Ph.D. proposal according to a certain federal agency guideline, then you would have learned that system as well. Two birds, one stone. Not that I advocate harming birds with stones!

Here is another reason why this is a good idea. In the 5 years after you graduate, federal agencies such as the National Aeronautics and Space Administration (NASA), Department of Defense (DOD), and the National Science Foundation (NSF) usually provide young investigators like yourself with five golden opportunities to write and win a proposal. Yes, five golden opportunities. If you strike gold (win) on the first try, you need not try again with this agency, although you could turn around and pitch another idea to another agency before those 5 years end. Get the message? So let's say, for example, that your work is especially suited for an agency such as NASA. NASA has the New Investigator Program (NIP). There are guidelines on how to write a NIP proposal including formats, page numbers, and the other guidelines. So let's pretend that you have actually graduated, and you are going to pitch your Ph.D. research idea and write a proposal. Well, this is exactly what my students do for one whopping credit in my Professional Development course. They write a proposal based on NASA's NIP guidelines.

My students pretend that they are writing a NASA NIP proposal, but in graduate school, they turn it into their committees. The master's-level students who take this course write a student fellowship to NASA Earth and Space Science Fellowships (NESSF) to fund their graduate school. The graduate advisors really like me because they do not have to labor over the various aspects of proposal writing with their students. So you've accomplished several goals in doing one thing.

Let's look at the alternative. You have to write a proposal for your committee, and you have no clue. You look at proposals that previous students have written, and they do not seem to match your specifications. Your advisor may know what she wants in a proposal but has trouble articulating it. Therefore, you have wasted a lot of time without a vision. Writing to federal guidelines will be required of you when you graduate, but with the NASA NIP method, you have learned a system.

I hope I have now made you happy and provided the purpose for writing a proposal. Well, purpose is one thing. Writing these darn proposals is a whole other story. Let me tell you this upfront. The first several proposals that you write are a painful experience. Why is this? Because you do not have a body of writing built up (in this case on your computer) that you can draw from (read as cut, paste, and massage!). But once you have written a few proposals, things start to get a bit easier. Be careful with the cut and paste though. There are plagiarism issues involved. I will address plagiarism in another chapter.

Now let me ask you something. Do you sit down, come up a great idea, write a proposal, and submit it? Looks right when I say it that way, doesn't it? Well, it's wrong. Let me explain why. If you did that in your committee, you may get away with it—maybe! But in the real world, this is not good practice for a winning proposal. Most proposals have a good bit of results already demonstrated in their proposal. I see your eyes rolling now! How can I get results before I have even started working on my idea? There's the first secret of writing a good proposal. If you wanted to increase your chances of winning a proposal in a competitive scenario, then you had better demonstrate "ahead of time" that you are capable. Why so?

As always in this book, we need to look at the other end of the spectrum. In this case, who will evaluate this proposal and under what metrics? Your proposal will be critiqued by several reviewers and/or panel members. Each reviewer may be reading many proposals. Then yours comes along with not only a brilliant idea but some demonstrated results. Compare this to only a brilliant idea from someone else. Whose proposal will get picked? No, you cannot go out and buy all copies of my book and hide it so your friends won't know the secret of putting results in proposals! Hey, buying a lot of books may not be a bad idea after all! My publisher will love it!

Let's quickly talk about proposal types. I realize that if I give a list of websites that request proposals or where you could pitch your ideas, this book could become quickly outdated. But, I will do it anyway, since I am all about practicality. Check out my Web resources (http://sachristopher.wordpress.com/) for various places you can write proposals to!

Types of proposals

Legitimately speaking, there are basically two kinds of proposals. The first type is when the agency (or organization) has a certain research problem that they are interested in, and they are looking to fund brilliant scientists, like you, to do the work and find solutions. Another kind of proposal is something like this: "Hey, we have a certain amount of money every year. We love research, and we have some broad mandates. Come look at our website, talk to us, write proposals, and do some work to further the good cause of science and humanity." The first is called a solicited proposal and the second, you guessed it—unsolicited.

Grants and contracts

Alright, now that you've got that, let's get this one out of the way as well. The difference between a grant and a contract! A grant is something like this: When an agency grants you money (technically, they do not grant YOU the money but the organization that you are working for), they want you to do good work (or you may not win a proposal the next time!), publish quality/quantity papers, and stay engaged in your arena of research. We'll talk about that later as well. You have a lot of freedom within the grant to explore a lot of exciting avenues above and beyond what you have proposed. A contract, as the word implies, is a bit painful though. Some scientists think that it curbs their intellectual freedom! When you win a contract, it means you have deliverables. It is like a ball and chain tied to your leg. You must state in the proposal what you will provide as a product or as a body of knowledge or whatever it is. You are held responsible for this until the very end!

Before you even think about your brilliant idea and how great it is, you must first spend a lot of time analyzing the request for proposal (RFP)—also called an announcement of opportunity (AO)—also called a solicitation—and a myriad of other things. Basically, it means that it is an opportunity to write a proposal. These documents are usually several pages long and provide specific instruction right down to how many pages a proposal should be and what font size to use. Since this is your first foray (unless you are a seasoned proposal-writing veteran who happened to get a copy of a book and wanted to be amused a bit), I'll make this statement. Read those instructions carefully and read them again and again until you fully understand all the requirements.

Now here's something that most people won't tell you. Go to the page that specifies how your proposal will be evaluated. On this page, you will find all the criteria as to how your proposal will be assessed by the reviewers. Usually, it has big sounding words like intellectual merit, responsiveness to announcement, relevance to the agencies, vision, cost realism, strength of team, etc. Sometimes they also tell you how many points they attribute to each one of those. Print that out, mark the key phrases carefully, and have it by your side when you write your proposal. Why? Because if these are the criteria according to which your proposal will be evaluated, you might as well fit your proposal into those categories. I hope this strategy makes sense.

I hope by now you've caught on that if you know what's expected and how your proposal will be evaluated, then you know what's needed. It is easier to write your proposals based on those criteria! Otherwise, it is like shooting in the dark! The same principle holds. Know what's needed and then work your way backward to fulfill those mandates!

But, first things first! Talk to the program manager! Who is a program manager? The program managers at an agency are the people who request proposals. They are actual people with real flesh and blood and real feelings. They make a living like most of us do. They are the all-important link between your proposal and the agency. They have had several years of experience working with scientists, budgets, and organizational needs, and it is a program manager who came up with the RFP for which you are writing this proposal. Too many people write a proposal without even consulting the program manager. Now the program managers can thank me for getting an increase in phone calls and e-mails because of my book! Nevertheless, it is vital that you contact the PM with your idea and get their feedback. Keep in mind that the program manager is not the one evaluating the scientific content of the proposal; that unpleasant task is left to the reviewers and panel members (more on that later). The PM will provide some details and guidance that may or may not have been written down in the RFP.

I suggest all new (and other) investigators first contact the PM via e-mail and set up a time to discuss the proposal idea. You may even e-mail your one-page summary or abstract to guide the conversation. In your conversation, briefly explain your idea and your thoughts on how your proposal fits into the RFP and the overall scope of the agency's program. Talk briefly about the innovativeness of your idea. It is quite OK to provide limitations and fallback strategies. Remember that this is a sales pitch!

The bottom line is to be upfront and don't undersell OR oversell your project! Remember that PMs have been immersed in this field (usually) for a while and are well aware of the real thing versus the smoke screen. After having explained your ideas briefly, ask for feedback and listen carefully to what they say about the proposal. Try to gauge whether they are enthusiastic or not. Most PM's will provide good guidance on what to emphasize and what to avoid. Take good notes. Ask questions if necessary. Thank the PM and also send a follow-up e-mail that you will indeed be submitting a proposal and taking into account their advice.

Budgets

Everyone likes money, so let's talk money.

Alright, the agency that you are going to write your proposal to should also tell you how many proposals they will fund and approximately how much money (usually a range) each proposal will be given each year. Most importantly, pay attention to the total number of years that this project will be funded. For example, the RFP might say that they expect to fund 20 proposals at $80–120K per year for 3 years. Just so that you know what 120K means, it is $120,000. You also need to have some idea as to when you can expect to get the money in your personal bank account. You wish! You don't really get money in your personal account; it is a complex billing procedure that your organization and the agency have in place. But we don't need to get into that.

Here's another tip. As a young investigator, do not expect to bring in multimillion-dollar proposals on your first or second attempt. The first few proposals that you win are usually considered seed money on the order of 80–120K per year. With the way things are going now, every proposal that most folks win looks like seed money. Once you have proved yourself, then, depending on the scope of the project, you can request more money. You'll hear this very often. Ask only for the money that you really need to do the project. If you only require 50K per project, then that's what you should request. Sounds righteous, doesn't it? I think so as well. But, I'll quickly show you why that amount buys little more than baked beans these days.

So let's make up a budget for you. This will be fun! Your RFP requires you to prepare a budget according to all the legal rules, and your Office of Sponsored Programs (OSP) is well aware of these rules. In case you did not catch it before, the OSP is those folks who tirelessly work to make sure that your proposal meets all the necessary guidelines. If it's anything like my university, they are understaffed and overworked. Treat the folks in the OSP properly; they could become your good friends (or in some cases, major irritants!). Let's assume that you are that same student from a previous chapter in the book, and you have landed your dream job at the university of your choice. They are only going to pay you 9 months in a year. Just to keep you eager, you have to go out and get those other 3 months of your salary. Read that as the carrot and stick paradigm! If your university is really vicious, they will only pay you 4.5 months per year, and you are responsible for the remaining 7.5 months!

The pressure is on, and since you like money and do not want to be without a summer salary, you are writing proposals like crazy. Assuming that you are successful, this could really backfire, especially if you don't have a mentor. You could end up with so many projects that you become quickly overwhelmed. That's another story for another book. But, I actually have a colleague who constantly finds himself in that situation. Too successful, too quickly. A recipe for a rapid burnout!

Let's assume that you have negotiated yourself a yearly salary of 60K. That sounds about reasonable. If ever you wanted to know about salaries, check out the University Corporation for Atmospheric Research's website. Plus, if you really want to know how much your professor makes, march on up to the library and ask. That's where they usually hide (er, keep) the salary information. Your professors are typically state employees, and most states by law have to provide this information. I can sense that I am getting deeper and deeper into trouble with some of my colleagues and peers with my advice! At 60K per year, your 9 month salary (academic year salary) is 45K, and your summer salary is 15K. This is a truckload more than what I made for my first job as an assistant professor, not too long ago. I was shortchanged! Moving on.

OK, you have 3 months of your time on this proposal. Is that a good or a bad thing? When you start out, you want to make sure that you're covered. Correct? You never know whether you will win this one or not. If you win this proposal, then your next 3 years are fully committed. A dilemma that we will sort out later! There are numerous bean counters these days making sure that when you say you worked 1 month on a proposal then that is indeed the case!

My junior postdocs begin to get really excited at this stage. They desperately want to ask for a lot of money so they can add not one but two graduate students to their proposal. Let's not get out of control now! Let's ask for one graduate student that you can advise AND mentor. Right? We are going to assume that since you have all your programming skills intact, you and one student should be able to pull off all of this work.

How much do you think a graduate student costs these days? A bundle! As a student writing this proposal, you will now step over the threshold into reality when you do budgets. I have a lot of fun doing this on the board. Let's pay the student $1500 a month in stipend. That's 18K per year. Well, now who pays the tuition, fees, and health insurance for the student? The proposal budget has to account for that. Each year that will cost you about, let's say, 5K.

You cannot help but glance at your salary and the yearly graduate student stipend and feel good that you are now making in 3 to 4 months what a graduate student makes in a whole year! But, who's got the responsibility? Remember my footloose and free statement earlier on in the book? The principal investigator (PI) has all the responsibility, not the student!

As a PI, you now need retirement benefits, health insurance, and fun stuff like that (these are called fringe benefits!). That's usually about 25–30% of your pay. So that will cost the project about 4K! You had no idea that you had to pay for these things, right? (This is similar to when I was younger. I had no idea what the fuss was about regarding electric/water bills for home and things like mortgage and rent, and then I grew up!) So you add up all of this, and your salary and fringe benefits are 42K. You are looking pretty excited since all of this is now coming together and everything looks doable!

I've already explained to you that you need to publish papers. Let's say that you are going to publish two papers per year. Well, the journals (most anyway) charge you to publish. After all, they have to typeset and maintain it on their website. They have staff as well. Note that I am a strong advocate of all these page charges, but it typically it costs about 2K per 10–12 page paper. That will run you about 4K per year. If you have a lot of color figures, then you have to shell out more money.

If you have a project, writing papers is not your only goal. You have to attend major conferences and meetings to present your work, so they will see the face behind the name. Plus it means networking. Check out the conferences section from the chapter on communication in this book. If you plan on traveling to two conferences a year, that will cost you about $1000 per trip. These are modest costs!

This will put your conference total at $2K per year. Remember that all of these are merely estimates. Some places are more expensive than others.

You may need miscellaneous things like books, software, and the like. You can add 1K to that miscellaneous category.

Add all that up and you get 50K. Now you are telling yourself that 50K is not a bad number after all since the goal was nearly 100K. I hate to break bad news right in the middle of all this fun, but there is such a thing called overhead. The politically correct term is facilities and administration (F&A) costs. Simply put, for you to be a PI in any organization, somebody is going to pay for your office, the light bills, the administrators (deans and VPs/president), and the support staff (administrative assistants and secretaries). Now you are thinking, wait a minute. Why should I pay MY project dollars for all of those folks? Taxes, taxes, taxes! Haven't you heard? Render unto Caesar what is rightfully his! Yes, I heard that. You capitalized that MY in the previous sentence! So when you hear your advisor talk about taxes, that is exactly what she means. You'll think differently when you become the dean someday. All of your proposal writing days will be behind you, and you may have not written a paper in ages. You will have to rely on PIs to bring in the money.

Anyway, you want to know what this F&A rate is. What percent of this 50K do you think it should be—10%? Go on, be generous—25%? Your project cost will go from 50K to nearly 63K at a 25% rate, so you are already screaming. But the F&A rate is actually about 50%. Yes, you have to live under these rules. F&A rates are different in different places, so in some cases it is even higher. Now you see that budgeting a proposal means that you have to include various items. At a 50% F&A rate, your 50K project will have an overhead of 25K and now your project will cost you 75K.

Wait a minute though! Now that you have a graduate student, you probably need a computer for him. Correct? Most agencies do not allow a PC as part of the project since word processing and spreadsheet type of work is not considered project worthy. Only real computing is allowed. So you have to buy a computing machine—not a word processing machine. Let's say that it is going to cost you 10K for a really modest computer with a UNIX/LINUX operating system, a good monitor, and lots of hard drive space. I do not advocate skimping on equipment. Provide a high-quality computing environment. Cheap stuff breaks often! Your grant total is now about 85K, which is a reasonable estimate. Note that in most cases equipment above a certain threshold, in this case 10K, is not taxed! It is well worth your while to research the rules of your particular organization to avoid unnecessary taxes.

Notice that taxes are levied on conference travel, publication costs, and everything except equipment over a certain threshold. There is no need to get upset at these taxes—that's the way the system works. If you do not like the system, you may want to think about starting your own company or laboratory.

So you will quickly come to the conclusion that as your salary goes up, your budget goes up. If you want more students, you cannot put them all in one proposal because this will put you over the 80K–120K range. Enter the rat race paradigm! This is why people write more than one proposal. Plus, these proposals are only for 3 years. Therefore, you need to plan ahead and write proposals regularly to sustain your research. While this may sound ominous, I take the glass half full approach rather than the glass half empty. Competitive proposal writing keeps your ideas new and gives you an incentive to publish and stay current.

I have provided a sample spreadsheet on the website (http://sachristopher. wordpress.com/) for you to play with. Table 2 shows how the budget usually looks. This is for year 1. See the Web resources for a sample budget for all 3 years. Since you need a pay raise annually, you have to account for that in the budget as well.

Table 2. Sample Budget for Year 1 Only

	Entry/Estimate
Organization name	University name goes here
Title of proposal	Title goes here
Period of performance	January 1, 2010 to December 31, 2012
Salary: Principal Investigator (3 months)	15K
Salary: Graduate student (1 year stipend)	18K
Graduate student tuition and benefits	5K
PI fringe benefits	4K
Journal publications (two per year)	4K
Conferences	2K
Miscellaneous (books and software)	2K
F&A Costs (based on 50K subjected to F&A)	25K
Computer	10K
Total cost	85K

Excursion

I've got to stop and tell this story. Every year, I have at least one or two students who really want the answer to this question: "So I write this proposal, and I win $300,000 over a 3 year period. Can I request whatever salary I want from that win?" This is why I love to teach the class. The innocence of it all! I gently tell them that their salary was set in stone by their organization when they got hired. Just because they won a proposal does not mean that they can now make 150K a year in the first and second years of their career. While salaries are linked to performance in most places, they are usually assessed by higher-ups in the food chain; yes, those that benefit from your F&A costs. No grinding your teeth now! (I am really hoping that your dean doesn't read this section!)

Here are some final thoughts on budgets. Contact the OSP soon after you know that you are going to write the proposal and get some help preparing the budget, getting it out of the way. This can ensure that the necessary signatures are obtained and the budget can proceed through the chain of command. Often, I see the young and the not-so-young investigators wait until the last minute to request a budget because, you guessed it right, poor time management. Look at it from another viewpoint, if you were working in the OSP, you would not want to see several proposals all at your desk for budgets and approval and other paperwork at the same time—would you? I would be unhappy with that PI from that point on! So give them a break. My OSP folks ought to definitely read this section. They will nod their heads vehemently. Well, I know what you are itching to ask, "How is Sundar managing time with his proposals?" I'll say this: I am one of the favored ones with my OSP team. Not for good looks, let me assure you of that, but because I turn in my proposals about a week before they are due (usually)!

The Actual Proposal

I hope you realize that you are not going to be the only one submitting a proposal for this RFP announcement. At least five to ten times the number of people who will be funded will apply. That means that well over a hundred young brilliant minds will be competing for that same pot of money. Now I hope this all makes sense that every edge you have—whether it is fine-tuning your proposal using already established results or implementing another of the various points I have raised—will affect your chances.

Before you start writing the actual proposal, here are a few pointers that are really worth their weight in gold!

Read the announcement or the RFP in great detail and then read it again until you really know the requirements. Make notes! How many proposals will they fund and at what level of funding? What are the areas in which they are soliciting the proposal? What is the fine print, the number of pages, and all the other requirements?

Make sure that your proposal idea fits the need of the agency soliciting the proposal. Sit down and come up with a one page summary that encompasses the following questions: What is the major goal of the proposal? Where and how does it fit the agency mandate? What are the specific objectives? What tools will I use (data, models, etc.). What are the expected outcomes? Who will benefit from this work and how? What are the broader impacts of the proposal?

The RFP may have other items that you need to address. Who are the investigators? Are there any collaborators? How much time will it take to do this project? Approximately how much money will you need and over what period of time?

Once you have formulated a simple plan for the technical and budgetary items, it is time to pick up the phone and talk to the all-important person: the program manager!

Before I go any further, I need to talk about how a proposal is selected so you will know how to write a proposal. As always, in this book, think about the outcome, and it becomes easier to perform the tasks to meet the desired outcomes.

Many RFPs are now online. This means that you need to know where to look for them. Check out my Web resources (http://sachristopher.wordpress.com/) for links to various agencies. Remember that some agencies solicit RFPs, and other agencies take proposals throughout the year. Therefore, you need to be aware of funding cycles. We'll talk about how RFPs are constructed at a later time.

Most agencies have two deadlines. A few weeks after an RFP is announced, a letter of intent (LOI) is requested. (See Web resources for a sample LOI.) The LOI does a few things. It alerts an agency as to how many proposals will be written.

However, there is no guarantee that those who submit an LOI will also submit a full proposal. An LOI is only about a page in length, and it highlights the major elements of the full proposal. A month or two after that (depending upon the agency), your full proposal will be due. This is the most painful part: the actual writing of the 15 to 20 page proposal.

Let's talk about what the actual proposal contains. Most RFPs whether NASA or NSF or NOAA will actually tell you how they want the proposal to be constructed. Whatever the subheadings might be, the common elements are the same. If it is an hypothesis-driven proposal, then you need to state that upfront and work the other elements in. Not all proposal ideas are conducive to hypothesis, but it should have a solid defendable scientific idea and a clear path from definition of the goals and objectives to solutions to the problem. With that in mind, here is a typical format for a proposal. (Check out the Web resources for actual winning proposals and links.)

Proposal Outline

1. Cover page (format may be prescribed)
   - Name/address/signature of a single Principal Investigator
   - Descriptive title of proposed investigation
   - Name(s)/address(es) of Co-Investigators (Co-Is)
   - Name/title/address/signature of authorizing institutional representative
2. Summary (abstract) of investigation (one page)
3. Budget summaries (total and for each year)
4. Table of contents
5. Scientific/technical section (not to exceed page limit):
   - Objectives and significance of proposed investigation
   - Technical approach and methodology of research
   - Relevance to agency (e.g., NASA) program(s)
   - Relationship to previous work in field and/or state of art
   - Management plan, including role of Co-Is
   - Timeline of activities (with key milestones as appropriate)
   - References/citations
6. Education/public outreach effort (if needed)
7. PI and Co-Is curriculum vitae (one page each)

8. Detailed budget and management plan (institution's own format, plus explanations)

9. Current and pending research support (for PI and key Co-Is)

10. Facilities and equipment (as appropriate, both for available and/or additional items to be purchased through grant)

11. Certifications—signed; required by U.S. Code (your institution will take care of this):
   • "Debarment, Suspension and Other Responsibility Matters"
   • "Drug Free Workplace"
   • "Lobbying" (only for budgets >$100K)

Although this list looks daunting at first, if you look very carefully, you will see that your work is primarily in section 5. This is where the crux of the information resides.

Before we talk about specifics of the various elements of the proposal, let me offer some pointers on how to split up the proposal into several sections and finish it in a timely manner.

Time management aspect of the proposal-writing process

Let's say that you have 90 days to prepare and write a proposal. This is usually a reasonable amount of time, especially if this is your first try at writing. Let me give you some guidelines on how to go about doing this. You need adequate time to write, review, and ask peers/colleagues for input. It is important that you ask (cajole) other proposal winners in your organization to review your proposal and provide critical feedback. If possible, ask colleagues who have won proposals from the agency you are writing to. This is why mentors are important!

You may think that 90 days is quite a bit of time, and you may want to put it off until the very last minute. Trust me; there is nothing more frustrating than having a good or a great idea and not allowing enough time to do a good job of proposal preparation. I've split the whole process into three monthly segments. To manage things effectively in each segment, I have further subdivided into three smaller pieces. Table 3 outlines how to allocate your time. Seasoned veterans who have the

Table 3. Sample Proposal Timeline

Segment 1	Segment 2	Segment 3
	Month 1	
Read RFP and read again.	Write relevance to agency based on RFP guidelines.	Write introduction with literature survey.
Create template with subheadings.	Work on hypothesis if applicable.	Write data section if applicable.
Make a checklist of all required elements of proposal.	Reach agreement with Co-Is about what sections they will write.	
Compile the literature survey.	Work with budget and work with Co-Is on their budgets.	
Refine outline with feedback from program manager.	Finalize budget with OSP and sign off on budget.	
	Write down overarching goal of proposed project and specific objectives.	
	Month 2	
Write methods and technical approach.	Create graphics: first cut.	Write project summary/abstract.
Have a clear path from objectives to methods.	Refine and finalize graphics.	Provide first draft to Co-Is and get approval.
Develop a clear plan of what you have done and what you want to do.	Write expected outcomes.	Typeset proposal carefully.
	Write management plan and budget justification.	
	Month 3	
Provide a clean copy to at least two internal reviewers in your organization who have been successful in winning proposals from that agency.	While waiting for their reviews, read your proposal objectively and write down critical issues. Are you explaining difficult concepts elegantly? Are you assuming too much? Evaluate your proposal from a reviewer's perspective.	Finalize proposal.
	After you get reviews, speak with your internal reviewers and then modify proposal accordingly	Go over checklist carefully.
		Upload proposal.

uncanny knack of putting together 15 page proposals in just a few days need not smile or snicker because we all have to start somewhere!

Remember that while you are thinking only about the technical portion of the proposal, there a variety of elements that an institution is responsible for. Budgets have to be signed by various heads, chairs, deans, and directors, and various templates required by federal agencies have to be signed and attested. The proposal needs to be checked for completeness to make sure that all rules are followed. Therefore, if a proposal is due on a Friday, do not e-mail or hand it in on Thursday. It will be met with chagrin by your OSP staff. Most institutions have a 72 hour policy. Your complete proposal must be e-mailed/uploaded or turned in to your organization 72 hours ahead of time. Otherwise, they will simply not process it. Therefore, make absolutely sure that you prepare accordingly.

With this in mind, here are some guidelines. Let's assume that the proposal is due on March 30. I usually suggest that you turn in the completed technical proposal to the OSP a week before the deadline. By doing this, you become one of their favorite PIs. Multiple proposals are submitted for each RFP, and by having yours ahead of time, they can process it and finish the submission process before all those procrastinators start bothering the OSP personnel.

What do I mean by "submitting it to the OSP?" Most proposals these days are all done online. All you have to do is to upload your technical portion to the website required by the federal agency and simply send an e-mail to the OSP in your organization stating that you are done.

If March 20 is the self-imposed deadline for you to submit to OSP and if you are looking at a 90 day window to get all of this done, you had better get cracking the very next day after New Year's Day. Remember that this proposal is not the only thing that you are doing. There may be other parts of your job—whether teaching, managing students, conducting research, or traveling—that interfere with this all-important task: writing the proposal. Now you know why that time management section I wrote comes in handy!

You are officially done at this point. Make sure that you check with the OSP to ensure that your proposal was submitted on time and request that they send you an e-mail that the proposal was done so you have official notification for your records. A few days after the agency receives the proposal, you will usually receive notification that they have received the proposal. The waiting begins!

Some pointers on various elements of the proposal

The abstract or the executive summary is the section that is usually written last because a lot of thought must go into this very short paragraph. It must clearly and concisely explain why this project is important, how it will fit the needs of the agency and the RFP, what the hypothesis is, and who will benefit from this. On top of that, key methods and data must be mentioned. Never write dull, boring abstracts. Try to capture the excitement of the project in this section.

The literature survey is very critical. You must demonstrate a thorough understanding of what the state of the art is and how you plan to contribute to the progress of the field. Remember that experts in the field will be reviewing the proposal. If you leave out key references and key research progress from a reviewer's portfolio, then you will not gain much momentum through the rest of the proposal. Therefore, do a thorough literature survey. If you have been reading papers diligently and documenting them, this should be easy.

If your proposal requires data processing of some sort, be realistic. You need to make sure that you carefully outline the amount and types of data and computer processing needs that fit with the overall scope of the project.

If your proposal requires building instruments, then make sure that the time allocated is realistic and that the RFP actually allows for this venture.

Make sure that you know the limitations of your data and methods and let the reviewer know upfront that you understand the issues well. Otherwise, the re-viewers may form the impression that you are unaware and therefore not suited to do this project.

Trying to be shy or modest is not the right thing to do when you are writing the proposal. If you have strengths in your team, then say so. If you have pioneered a certain method, then say so. However, don't get too carried away with words or be heavy-handed.

Try to have one overall goal and several specific objectives that will allow you to reach that goal. Carefully divide the time allocated for the project into those goals. Again, be realistic.

If you have a hypothesis-driven proposal, that is an added bonus for some agencies. Make sure that the hypothesis is meaningful.

The methods must be clearly explained. If there are major flaws in your methods and approach, then it is next to impossible to win a proposal. Do not assume that the reviewer is an expert in that field. Learn how to take complex concepts and briefly explain them. Provide references for additional material. A proposal should read like a story. A great start, some healthy tension, and a great ending!

Figures must be top-notch with good labels and crisp axes and numbers. If color is required (often a good idea), then make sure that you use appropriate colors. Yes, you can use a figure from elsewhere to explain a concept, but I'd rather have you come up with a figure. Remember one of the goals of the proposal is to judge your capabilities. Equations are fine, but don't throw in equations just to impress the reviewer. Clearly explain all terms and break down the physical significance of the equation.

Spend some time on a management plan and a timeline. How long will it take to complete a project? Have contingency plans. Tell the reviewer what the responsibilities of each person in the proposed project are.

CVs must only be one or two pages. Therefore, write down relevant information and relevant papers. It will certainly impress the reviewer if you have already published a lot of papers.

All researchers are now asked to think of the broader impacts or education and outreach elements of their work. It is important for you to note how your proposal will benefit others in the field or the larger community.

Top Ten Dos for Proposal Writing

1. Carefully read and follow the instructions in the RFP (request for proposal) of interest.
2. Focus the proposal on a single, clearly stated scientific problem and describe its research plan and anticipated results. (Clearly address the "so what?" and "why do this?" questions. It's equally important to show how the proposed investigation will advance knowledge in the field.)
3. Provide judicious amounts of tutorial material, especially if proposing innovative work. (Not everyone reviewing the proposal can or will be an expert in all aspects of the proposed research.)

4. Give credit to others as appropriate by including references to preceding work in the field.

5. Proofread and spell check the proposal before submission. If at all possible, have a colleague review it. (Strive for a quality of text commensurate with an article prepared for submission to a journal.)

6. Keep the proposal text as short as possible consistent with completeness. (Page limits are "limits" not quotas to be filled.)

7. Use clear, easy to read type fonts and page layouts of material.

8. Include all requested information. (It is all used in the review/selection/ funding cycle.)

9. Strive for realism as well as adequacy of the requested budget.

10. Provide any and all material necessary to understand the budget.

Proposal Review Process

Once a proposal is submitted, various things happen within an agency depending upon protocol. The agencies have to find reviewers who are experts in your field of interest. A proposal could be reviewed from three to five reviewers or sometimes even more. The program manager in charge of the proposals is responsible for securing the appropriate reviewers. Now I hope it makes sense why it is imperative for you to speak with the program manager BEFORE you even write a proposal. After all the reviews have been received for all proposals, the PM has the unpleasant task of the selection process, picking the winners of the proposals. It's unpleasant because if there are winners, then, by necessity, some won't win!

At this point in time it is important—very important—for you to know the criteria that the reviewer used to rate your proposal. Ask no further. Here are the metrics. Usually, they are all spelled out in the RFP. If you have done your homework properly, you should already know what the rating criteria are. A sample evaluation form that reviewers typically use is provided as a Web resource on my site. In general, you are rated on (1) relevance to the agencies mission and mandate, (2) technical merit, (3) the investigators' expertise and strength, (4) the realism of costs proposed, and (5) an overall evaluation. The rating for each criteria ranges from excellent (E) that usually gets five points, very good (VG) (4 points), good (G) (3 points), fair (F) (2 points), poor (P) (1 point), and high risk (HR), which is often debatable. More on that later! The overall rating is crucial for the proposal since overall impressions are marked by each reviewer.

By now, you should have guessed that this is not an entirely flawless system. Points and ratings are subjective. I also hope that you have caught on that every little thing counts when putting together a proposal since the competition is intense. I hate to say this, but while some are cynical of this whole process of proposal writing, selections, and such, no one has ever come up with a better alternative. I guess you could stand on a street corner in Washington, D. C., somewhere close to the agency and scream at the top of your lungs that you are a good researcher and you need money to do research.

For the most part, the system is absolutely fair since the agencies ensure that every proposal has an equal chance of winning. Guaranteed, there are some excellent proposals that are not funded and for that matter a few poor ones that do get funded, but a well-written proposal usually gets funded. Optimist that I am!

The PM now has the formidable task of looking at ratings, the amount of money available, and all the associated metrics to make a choice of funding N number of proposals set forth in the RFP. For this, she usually relies on peers at the PM level or a set of panel members who are invited to review and discuss the metrics of the proposal. Note that the reviewers in the first stage had no clue who the other reviewers were and simply read the proposal and rated it based on their scientific expertise and submitted it to the PM.

The panel members are a superset of reviewers. Usually, about a half dozen to a dozen are invited to discuss the proposals and the reviews to help guide the PM to select the best proposals. Briefly, each proposal is presented by a lead panel member and one or two coleads followed by an open discussion and a rating of E, VG, G, F, or P. Each proposal is presented over a 2 to 3 day period, debated, discussed, and rated to help the PM fully understand any elements of the proposals that may have been missed at the review stage. On the last day—based on the panel discussion—the PM rates the top proposals by assigning them to various categories. If the panel gave a proposal an E rating, there is nothing much to debate. If it is a poor proposal, then there is no debate either. Armed with a list of proposals, the PM in conjunction with his upper level manager will make final decisions on whether a proposal should be funded.

I say all of this to provide some concrete pointers for writing the proposal. When you understand the full process, then you have a good sense of how to forge a winning proposal. The process involving some combination of reviewers and panel members is used in most agencies to make decisions. See, I told you—it is a fair system!

The envelope, please.

As a young investigator, you wait with bated breath to find out the results of the proposal. After all, you have spent weeks reading lots of papers and other documents, writing, wordsmithing, and refining your proposal. Obviously, this is like your own child. Your child is the prettiest, the strongest, the fastest, and the best. You cannot bear to hear anyone even remotely mention flaws in your child. Translate this to your proposal; it is the prettiest, the best, etc. Therein is the problem. We tend to take proposal writing personally. We'll talk about scenario 1.

For the euphoric: The happy letter!

You open your sealed, signed letter from the agency and read: "Dear Dr. Eastwood, We are pleased to inform you that we have selected you proposal for funding." The letter goes on for another two pages, but you have already stopped reading. You first close your office door (if you have your own office) and literally dance and leap for joy. Next, you walk the corridors beaming—hoping that someone will ask you about the proposal that you wrote a while ago. Of course, no one does, and you cannot wait to break the news to your colleagues and friends and your boss. You start every sentence for the next day or two by saying, "Guess what I heard yesterday," and then you go on to let folks know about the winning proposal.

OK, I exaggerate!

After you get back to planet Earth, I strongly recommend that you read your acceptance letter carefully and make sure that you follow the instructions outlined in that letter. Often, the reviews are enclosed. If not, I suggest that you write to your PM and request the evaluations. These reviews are priceless. Often, they provide new ideas that you can explore within your project. So read and assimilate the information carefully. Make sure that your OSP personnel know about this. I have already written quite a bit about proposals and the selection process, and I am not going to talk about how to manage a project here at this stage. You'll have to wait for my next book! Read the letter to see if you need to contact the PM or if other steps are required. Either way, congratulations, you have just won your first proposal!

For the downcast: The sad letter!

Let's talk about scenario 2. Your letter states that your proposal has not been selected for funding. Obviously, it is a disappointment. Trust me; I have had numerous letters of rejection from various agencies in my career! While it is never easy to get one of these letters, it is absolutely important that you stay focused and not get cynical about the process. I often tell my students to simply put the letter down and not get back to it for a couple days until all rationality has returned.

Now you can read the reviews carefully. Yes, I know that all the reviews are bogus because you did not get selected, right? Not really. Oftentimes, there are some valuable points that the reviewers have raised. Maybe the proposal had some technical shortcomings. Usually, I have found that it was not a case of short-comings, but it was failure on my part—I did not explain or reference something adequately. As a PI, sometimes I tend to assume that the reviewer "should know" about a certain aspect when, in fact, I should have taken the time to write the principles or the methods better. Either way, learn not to focus on the reviewer but on how you could craft the proposal better the next time.

I have often found this with my colleagues. If cynicism and disillusionment sets in, then the proposal never goes anywhere. However, if you truly think that it is a good idea worthy of funding, you can do something about it. First, after reading the reviews, go back and read your proposal. Make notes on your proposal as to what you would change if you could be given a second chance. Now you really can believe in yourself.

The best piece of advice that I can give you is this. Work a bit more on your ideas. Maybe you can even write a short paper on one aspect of your proposal. Rewrite your proposal by taking the reviewer comments into account. Resubmit the proposal. Keep in mind that there are never ever guarantees that a proposal will be funded the second or the third time. But with each turn, you are simply increasing your chances for success. If you did not submit a proposal, then the chances of you winning are zero!

Whether you are funded or not through this iterative sequence, you are the beneficiary because you have learned how to write and communicate better each time and hopefully published some good papers along the way.

> **Bottom line:** It is too easy to get disillusioned or downcast. Don't! Concentrate on what you can do and not on the agency or the reviewers.

Proposal Evaluation

Evaluate the proposals that you are given based on the criteria given below. Carefully review each proposal and provide appropriate comments.

> *Note to instructors:* This exercise will certainly depend upon the number of students you have in class. Each student should be given the opportunity to review proposals of their peers making sure that is anonymous. For example student A, who may be reviewing the proposals of student B and student C, should never divulge that he is reviewing them. He, however, knows whose proposals he is reviewing.

The instructor should hand out the evaluation form shown in Figure 1. Once the instructor receives all the reviews, she should go through a simple rating system similar to a real-world scenario. Pick the best three proposals (as an example) and announce the selection list. Include a panel discussion for each proposal if time permits!

Proposal Evaluation Form (This is a template)

Proposal:

PI Name:

Institution:

Title of proposal:

Evaluation Guidelines

E (excellent):   Outstanding: presents an opportunity for a major contribution to our understanding. I urge support.

VG (very good):   Important contribution. I recommend support.
G (good):   Competent: will make a contribution, but the proposal is routine or has correctable deficiencies. I suggest support if funds are available.
F (fair):   Satisfactory in part: possibility for a contribution appears limited; routine in character. I suggest rejection in its present form.
P (poor):   Unsatisfactory. I recommend rejection.
HR (high risk):   Scientifically intriguing, innovative, and addresses important problem. Difficult to evaluate because of reasons lists under "additional comments." I recommend support on a trial basis if funds are available.

Relevance to Agency Objectives (Circle One)
1. Potential Contribution to Agencies Mission E VG G F P HR
Strengths:
Weaknesses:

Proposal Intrinsic Merit
1. Scientific Merit E VG G F P HR
Strengths:
Weaknesses:

2. **Offeror's Contributions** E VG G F P HR
Strengths:
Weaknesses:

3. **Personnel** E VG G F P HR
Strengths:
Weaknesses:

Cost
1. Cost Realism E VG G F P HR
Strengths:
Weaknesses:

OVERALL EVALUATION E VG G F P HR
Overall Strengths:
Overall Weaknesses:

Suggestions for Proposers:

Evaluator agrees to use the information contained in this proposal only for the purpose of performing this evaluation and to return all copies and associated data upon completion of the evaluation.

Evaluator's Signature _____

Evaluator's Name (typed): _____ Date _____

Figure 1. Sample proposal evaluation.

Notes

8

Communicating

Learn to Talk

Be excellent in speech (and all forms of communication).

You must learn how to write and speak effectively. You have to do this over and over again, especially if you get a graduate degree. There are no two ways about it. We'll talk about writing later, but let's tackle oral communication first. Needless to say, numerous books have been written, and good communicators make a lot of money teaching other people how to communicate. The so-called gift of gab is not entirely a trait that you are born with—it can be developed!

Conferences are great places to learn about communication styles. The range is quite diverse—from the slick and suave to the downright boring. So what makes some people good speakers, while others never seem to get beyond that mediocre threshold? I believe that anyone can improve with sufficient practice and with deliberate adjustments to communication styles. That's the key. Even when you think you have arrived at giving great talks, find ways to get better. I say this from experience since even some of the "giants" in the field always strive to make things clearer and communicate more effectively.

Before I give you some pointers on some of the annoying things to avoid, let me briefly say that the key to any talk is preparation: knowing your material well. In most situations, you have 15 minutes or less to speak and answer questions. So you have to prepare and rehearse well and make sure that the take-home messages from your talk are clear. Most people will remember only one or two major points from your talk, so make sure that you emphasize those points. Knowing your audience is equally important. Stick to the allotted time because you do not want the moderator to cut you off when you are only halfway through your talk. Careful rehearsing is necessary. Usually you speak for the allotted time, and the questions are at the very end. Therefore, practice your talks well until you become an expert at managing the time factor in your talks. As you progress through your career, you will practice less, but you have the thing called experience under your belt! Until then, practice!

Here are some pointers on presenting your research in 15 to 20 minutes. I will save the topic of how to teach a 50 to 80 minute class for a later time or another book!

If you struggle with the fascinating language called English, then take some classes on communication. I've even suggested that my foreign-born students whose native tongue is not English fast from their mother tongue for a year and speak exclusively in English. I have asked them to look for an English-speaking roommate rather than someone from their own culture. Forced methods like this quickly bear fruit. But only for the brave!

Know your audience. This sounds obvious, doesn't it? I once prepared a long talk that had great slides and fantastic technical material. When I got to the venue and started speaking, I realized that very few in the audience had any idea about the topical area that I was discussing. I forgot one minor detail before I started making my slides: to ask the organizer what the audience level was. Needless to say, it was a terrible talk. I had to skip slides because most of the material I had prepared was way overboard!

Make eye contact. Have you ever been in a presentation where the speaker refused to make eye contact? It looks dodgy! Everyone feels uneasy when the speaker looks only at the slides or a white board or, worse still, at a pointer or their shoes. Eye contact is absolutely crucial if you want your talk to feel good! Note that I said "feel good." Talks are effective when the content and the "feel good" factor are correct.

Be careful not to keep your hands in your pockets or wave excessively. This only works in a magic show.

Be careful not to speak too quickly. It's no fun for you or your audience when you spew words at a thousand words per minute. It sounds like subway chatter.

Go easy on the tables and numbers. You may love your tables with 3000 numbers but the audience could care less. In your eagerness to impress, don't make your audience dizzy with too many equations with a lot of Greek symbols! You need to develop the art of taking difficult things and explaining them in simple and elegant ways. That's why it is important to attend seminars as a graduate student and learn from diverse speakers. This will help you make the transition from an average speaker to a great one!

Use legible slides. I am not going to give you pointers such as three bullets per slide or types of font and colors except to caution you to avoid yellow letters on a yellow background. The audience in the back will NOT move their chairs to the front to read your slides. Common sense works when it comes to the number of figures, bullets, and items on a slide. Pick a format and a comfortable background, then keep things uniform on each slide. After you finish creating the slides, put them up in the presentation room and walk to the back of the room. Make sure that you'd want to listen to yourself with those slides. Don't make your slides too wordy since the audience is there to listen to you, not read from your slides. If that were the case, they would prefer going to a local coffee shop or the library.

Don't get too big for your britches. Tell your audience your outline and make sure you include a slide with your title and name.

Go easy on humor. We are not paid to be stand-up comics (or are we?). On that note, nothing is more annoying than sounds during slide transitions and obnoxious noises embedded in your presentation. Have you ever seen a presentation that ended with a huge thank you slide and an applause sound clip? There's a limit!

During the question and answer session, wait until the question is finished, repeat the question slowly so everyone can hear and then answer the question in sentences not in monosyllables. Repeating the question is a good habit since oftentimes the person asking the question is in one corner of the room with no microphone.

As your communication skills evolve, strive to engage your audience with your words and the slides. That is the most important skill. If all else fails, take a few classes on how to communicate better.

As a graduate student, you have a unique vantage point. You listen to a lot of different instructors when you take classes. If your department is doing things right, you are also listening to an outside speaker every week or month in some form of seminar series. While some students may think it is a waste of time to attend so many seminars, use this time slot to observe communication habits and skills.

Watching others deliver a lecture gives you a lot of insight on how or how not to present your material when the occasion arises. Believe me, it is easy to criticize someone about the way they deliver a lecture or a seminar, but when you get in front of crowd, things look different. It is your turn to be critiqued!

Practice. If you were like me, you tried your best to avoid every speaking opportunity. I regret it now! Have you ever heard an adult say with a remorseful voice, "I wish I had listened to my mom and stuck with my piano classes." Well, this remorse is the same, since no one ever told me what I am telling you now. Volunteer to give talks while in graduate school. Hone your communication skills while in graduate school. Look at it this way. Your audience in graduate school is a whole lot kinder than the outside conference and science team audiences. So let loose and practice.

Most foreign-born nationals who did not speak much English in their country have a tough time speaking in front of an audience. The same goes for some of the students born on local soil as well. My strongest advice is to put yourself in situations that require you to speak. Volunteer to give a seminar once a semester. Ask for feedback after every speaking engagement and watch your confidence grow!

I've got to tell this story about communication styles. An eminent researcher came through one of our seminar series a while ago during the days when we were just making the transition from slide/transparency projector to PowerPoint. There were at least a hundred people in the audience. The first transparency did not have a title or any explanation, just a graphic of some sort. For the next 10 minutes (10 minutes that seemed like an eternity), he paced back and forth IN FRONT of the projector with the pointer tucked underneath his chin staring at his shoes. All the time he created shadow and light intermittently from the overhead projector. Yes, in front of the projector with the light on! The audience was shocked and thoroughly bored, but none had the nerve to do anything about it. I bet that most students in that room made a mental note to never pace in front of an overhead slide for 10 minutes. I am sure that you have a lot of stories from your end about bad communication habits as well. Every time I teach my Professional Development class, I make sure that I reenact this situation!

I think I've beaten the oral communication horse quite a bit now. Let's briefly talk about writing skills. I have yet to find a student/researcher/professor who claims to actually like writing, especially formal writing. I don't. Confession aside, formal writing is painful. (I would not call this book formal by any stretch of imagination, and my editor nods vehemently!) You think that your term papers or your project papers were bad? Wait until you write your first peer-reviewed journal paper.

Let me recount one of my first ever writing experiences. I still remember my first foray into journal paper writing. I took a lot of time, which undoubtedly annoyed my advisor in the process. I analyzed and reanalyzed the data to make sure that things were right. I was just getting ready to tell my advisor that I needed more data when he came bearing down, in a nice kind of way, and demanded that I give him that first draft. I had already written that first draft and gone through it several times. He took it home (probably because surgery with a red-ink pen is best performed without interruptions). I got the draft back in about 3 days or so. I had never seen 20 pieces of white paper with black type with so much red on them. It looked like the whole thing was written in red. There were so many red marks all over that first draft that I had no clue where to start repairing that document. Sure, I was devastated, but I knew that this was necessary to learn how to write effectively. My subsequent drafts and papers through this same advisor improved; there was less and less red ink. It's funny that my students go through the same process now. For most, writing never becomes fun, but it is necessary. Without it, it is difficult to survive. You learn how to develop a thick skin—a necessary trait for any budding graduate!

How does one get better at writing? There are no magic bullets here. It's just repeated, deliberate practice. Does this sound like the oral communication section? Right on! That's the truth though. There are a lot of books available on how to write effectively. Honestly, as a graduate student and as a professional, I find those books hard to read and even harder to implement in my work. Therefore, when I read papers, I make mental notes on some of the better ways of communicating a sentence. Pattern yourself after a good writer. It may or may not be your advisor. Nevertheless, find a pattern, work through it, and be open to constructive criticism. I am not implying that poor grammar skills can be overlooked. You need to have a good idea of how to construct a sentence properly and then work on it over and over again. True, some can make a sentence read better than others, but if you are communicating effectively, I wouldn't worry too much. With time, your writing will improve.

Often when an advisor corrects a student, the student never stops to look at that corrected to sentence to learn from his mistake. He simply accepts the track changes correction and moves on. This is another example where word processing software is not a good tool for a student who is learning how to write. If you were to pay attention to why that correction was made, you probably will be more aware the next time around and less likely to repeat it!

Writing opportunities abound. Remember that one of the beginning take-home messages was to maintain a journal. Writing in your journal on a routine basis will help improve your writing. Also, when you make a figure or a table of some sort for your research, spend a few minutes to write in formal terms what that graphic is telling you. Blog it! By the time this book comes out, blogging may be a thing of the past. We may have moved on to the next thing in communication via the Web! Nevertheless, use a tool—any tool—and be open to criticism!

This is solid practice and provides much needed growth. If you do this routinely, you are creating valuable material that you can use later when you actually start writing your papers/thesis. This would be a case where a word processor is useful. Once you have gathered a lot of material, you can cut and paste at a later date.

Effective writing skills are a lifelong venture. Learn from mistakes and be a disciplined, consistent writer.

> **Useful tip:** Students may not know that most universities have a language center to provide help with writing—free of charge. Tutors help students one-on-one with their writing. Use this service to improve your writing!

Journal Entry

What are the topics for your seminars this month and this year? Purpose to improve your speaking skills.

Food for thought

1. Perform a strength and weakness analysis on your communication skills. Ask your advisor to discuss your strengths and weakness as well.
2. List specific things that you could do to minimize those weak areas in your communication arsenal.
3. As a graduate student, what types of opportunities are available to you to communicate?
4. Draw up a 1 year plan to put yourself in situations that will require communicating in formal settings.

Conferences

Do not hide under a bushel.

Ah! The word conference conjures up images of traveling to Hawaii and listening to talks in the morning and snorkeling in the evening after the conference. Well, let me be the first to burst that bubble. Conferences are not all that glamorous, and they usually require a lot of work and preparation. They can be downright draining.

In most disciplines, there are the revered and the most anticipated annual or biannual meetings. They give you an opportunity to get out of town, travel by plane (usually), stay in a hotel, eat in strange restaurants, meet your colleagues, make new friends, network, listen to interesting talks (and some boring ones), eat dinner while talking science, and lay awake in bed in a strange hotel in spite of the 17 pillows they have to fit your neck curvature just to get up exhausted. And you do this for 3 to 5 days! Cynicism aside, conferences can be good or bad. Some conferences last for 5 days, and there are at least 50 sessions throughout the week. Even your electronic organizer cannot keep up with the scheduling of the talks.

Needless to say, it is a nightmare! Other conferences are smaller and are much more focused, so you usually stay in one or two conference rooms the entire time. I tend to like these because I can learn more from these focused meetings.

Here is how the path to a conference works. Conferences are announced months or even a year in advance. The topical areas are listed on the website, and you are asked to submit an abstract (less than a page) and indicate whether you would like to present it orally or in the form of a poster. While most folks like to "talk" since it is considered prestigious, I believe that a poster can be equally important. So I suggest that for every conference that you go to (if permitted by the organizers), ask for one poster and for one paper presentation. It is a tremendous experience to interact both ways. There is no guarantee that even if you asked for an oral presentation that you may be given one. You may end up with two posters. It's not that the organizers do not like you; they are simply juggling several things to accommodate posters and presentations alike. Since hundreds of scientists write such abstracts, the organizers have to group them. Depending upon priority, they allocate some for talks and a vast majority for posters. Some conferences only require you to write a one page abstract. Others either require you to write an extended abstract, usually three to four pages, or leave it up to you to write one or not.

A conference abstract is usually written months before you even attend and present at the conference. Therefore, you are doing some work between the time of abstract submission deadline and the actual conference itself. Apart from the obvious—title, author list, and affiliations—remember to specify the session you are interested in and whether or not you are planning an oral or poster presentation. The abstract should highlight why this work is important, what some of the data sets and methods are, some results that you already have, and what you are working on. After you submit your abstract, make sure that when you travel to the conference, you have the same title. Switching titles and focus of the paper is unfair to the audience because in a big conference they have juggled schedules to get to your talk. You need to make sure that you are presenting something true to your abstract.

Imagine this: You are at a large annual conference in California just before Christmas, and there are multiple parallel sessions. All you have is a conference proceedings book that you were given when you registered. You look through the abstracts, and you are excited because someone is going talk about the effect of agriculture on air quality at 11:30 A.M. in a certain room. You have another talk that you would also like to attend, but you make sure that you there at 11:30 to hear about air pollution. The abstract had looked rather interesting since that is your area of research. At 11:30, if the speaker stands up and starts talking about some field experiment results related to water vapor, this can be hugely disappointing to you. Therefore, when you write your abstract, make sure that you will talk about things that you say you will when you wrote the abstract.

I was once asked to come up with five things a student should do at a conference to organize their time. Remember that there are a lot of things going on with talks happening in several rooms all at the same time. How does one formulate a plan? I suggest dividing your time between these five parts. Obviously, the percentage of time in each of these sections will vary depending upon where you are in graduate school. Here are some guidelines:

1. Attend talks that will strengthen your specific research topics.
2. Attend talks that will broaden and enhance your research.
3. Learn how to have fun at the conference venue.
4. Interact with peers from other universities and organizations. Networking with your peers pays huge dividends. Peers today, leaders tomorrow!
5. Finally, set aside some time to talk to potential mentors and some of the icons of the field. Most senior researchers enjoy interacting with graduate students.

There are several reasons for attending a conference. This is an opportunity for the world to see the face behind the name—you! This is why giving good talks is important. Most talks last 12 minutes, yes, 12 minutes with about 3 minutes for questions. Therefore, claim your fame during your talk!

Conferences are also for networking. Networking simply means that you meet people of similar research interests so that you can collaborate with them to further your research and, get this, theirs! Conferences are also great places to meet potential mentors and to gain ideas for deepening or diversifying research.

Here's something you won't hear in any book (except this one!). If you do not plan on interacting at a conference, don't even go. If all you will do is interact with your office mates or your friends from your workplace, you might as well do that over a cup of hot chocolate at a coffee shop in your neighborhood rather than travel all the way to a conference. It's a waste of time and money. At the end of each conference, you should be able to sit down and list specifically the people with whom you interacted and the follow-on action items as a result of those interactions.

This is also a great opportunity to seek out those in the field who have written papers and conducted field experiments. So muster up courage and introduce yourself. Most senior scientists welcome the young researcher in the field and are eager provide great advice. Early in my career, at an annual conference, I got up to give my talk and saw one of the leading researchers in the field in the audience. He asked a difficult question, and I thought I answered it reasonably well although I wasn't sure. During the break, I approached him and introduced myself. He spoke with me for a while, encouraged me, and provided some valuable advice on how to move my research forward. As a parting thought, he said that he welcomed my collaboration and indicated that there was lots of room in that field. It left a lasting impression that I carry with me today. Even though he passed away prematurely in an accident recently, we wrote several papers together, and I learned from him every single time! Remember that there is a lot of room for people to grow together in your field of interest. Do not slam the door shut on people. They are not competitors but collaborators!

> **A note about answering questions:** Have you ever been at a conference and at the end of a talk, a hand goes up on one side of the room? A question is raised that you can hardly hear, yet the speaker starts answering the question. None except the few people around the person who asked the question have heard the question. Another annoying aspect is when the speaker does not wait for the question to be finished but starts answering in the middle of the question. Rude, wouldn't you say? Or, my favorite, the rambling question that takes up almost a minute or the rambling answer that is not even close to the question that was raised. The list of negatives could be endless.

With that in mind, here are some thoughts. Wait until the question has been completed. Repeat the question if the audience did not get an opportunity to hear it. Answer the question concisely and clearly without being condescending.

If you are attending a conference, then you should present a paper or two. Let me say this upfront even though I will get into trouble for this. Conference papers that appear in proceedings usually count for nothing on your resume as far as demonstrating your productivity simply because they are not peer reviewed. So do not waste your time on writing conference papers unless it is for practice!

Do not waste your time at conferences by endlessly wandering the corridors. It is like a theme park. I didn't say a zoo, but a theme park! You have a short period of time, and if you want to see all the attractions, you had better do some preparation before you land at the conference. Check out the conference proceedings online before you go. Mark all the talks that interest you. Jot down all the folks you would like to meet and interact with and follow a game plan. While you are at it make sure that you have fun after the conference ends or in the evenings, getting to know people beyond only the scientific realm. Lasting relationships are formed during these conferences.

I still remember going to South America during the first year of my life as an assistant professor for a conference. I met a young and energetic graduate student at that time. Little did I know that he would run major field campaigns and become a major force in the field! We ate a lot of meals together and talked about the good and bad of science. We are now good friends both on and off the scientific field. You never know what might happen to the person with whom you interact today.

One last piece of advice! Don't travel too much to attend conferences and only present papers. While it may be good for a while, it does take valuable time away from doing research and that bottom line: Writing those papers! Be sensible about allocating time for conferences and making the best use of your time.

Journal Entry

I like conferences because _____

I hate conferences because _____

Food for thought

1. Assume that your first conference is coming up. How will you allocate your time for an effective experience?
2. What are your strategies for networking among peers and senior researchers?
3. What is your plan of action for follow-up with peers and researchers after you get back?
4. Write down your goals for this conference and how you will benefit from this experience.
5. If you gave a talk at this conference, what feedback did you solicit and how will you improve your talk the next time around.
6. If you presented a poster, evaluate your feedback and how you could improve a future poster.

Notes

9

Teams

Pick Your Team

In the counsel of many, there is wisdom!

Beyond just your advisor, you must think, or learn to think, of your CMs as part of your team. Yes, you heard that right, your team. CMs provide balance and valuable advice on research and guidance on career-related matters, provided that you ask them. Most often committees are put together hurriedly before an exam or a thesis/dissertation defense to adhere to university guidelines. Big mistake!

In graduate school, you are expected to write a thesis and/or defend your research in front of your advisor and committee members. Having CMs ensures that you don't get off easy (just kidding!). It is a time-honored graduate school tradition that there are three to five members in your committee. Often these members are from within the department, and I explain later as to why that this is poor practice. Your research must have a national and sometimes international perspective. Therefore, it is important to pick at least one or two members from outside the department.

Unfortunately, the student often picks CMs based on how "soft" that particular member might be. What do I mean by that? At the graduate level, if you are a master's student, the CM is often asked to provide advice on the thesis. For a Ph.D. student, the CM is involved in writing and asking questions ("grilling" as it is affectionately called) for preliminary and qualifying exam questions in addition to providing guidance on research. There is a notion that you need to avoid CMs who are traditionally rigorous in their questioning and in the exams. I think it is a serious mistake to go that route. Unfortunately, sometimes students like to take the path of least resistance by requesting CMs who have a reputation for asking easy questions. Remember that as a graduate student this is one of the best opportunities to work with leaders in your area of interest. Therefore, the best CM for the topical area must be sought after. Remember that the CMs are on your side. They want you to pass, do good research, and contribute to the good reputation of the university. So pick the best team possible and learn how to appreciate the different vantage points that come with various CMs.

I strongly suggest that the student and the advisor take a deliberate approach when picking CMs. First, check the graduate school guidelines to make sure that the makeup of your committee will satisfy the requirements of the graduate school. There are concrete rules in place in terms of who can serve on the committee and the number of members on the committee. Stay within the rule-book. Then you must determine the scope of your research. Survey the department, the university, and the committee. Here is a huge tip. If you are really interested in working for a particular group or person at a certain organization after graduation, it may be in your best interest to get them on your committee. This is a good pathway to that dream job of yours. Pretty nifty strategy, wouldn't you say! I routinely try to have at least one member from a major federal agency or lab on my student's committee so the research results get out the doors of the department even before the student defends. Picking such a committee member provides a unique, outside-the-box perspective, and they can be an ally in the real world for the student, providing a potential job offer for the student after graduation—a win-win for all.

If rules and regulations require you to have three out of five CMs from your department, I am sure that you can ask for a sixth member if it will help advance your career goals. Check with the graduate school dean or coordinator.

While picking your committee, you have full say in the matter, but make sure that you consult your advisor. He knows about conflicting issues between and among faculty members and groups, and he can steer you away from potentially dangerous committee meetings and thesis/dissertation defenses. I know of ugly Ph.D. meetings when CMs pull out all the stops and engage in an ugly war on words while the student is left bewildered! Play it safe. Check with your advisor before you pick your CMs.

Let me share some advice as well as some guidance about what your CMs expect from you. The committee member, while part of your team, usually makes the assumption that you are working closely with your advisor. If you are doing all the things I have outlined in the previous chapter, then that is indeed the case. You are managing your advisor well, but no need to tell him that!

Next, the CM assumes that you will contact her at periodic intervals to provide updates and reports on how things are progressing. Trust me. No CM wants to come in cold to your defense and see for the first time results that make no sense or results that have not been given any feedback. So it is in your absolute best interest to manage this aspect of your research. Here is a priceless tip. Set a target date for your defense. Then mark on your calendar when the first draft is due to your advisor and the committee. On that same calendar, mark approximate dates when you will meet with each CM separately to provide a progress report. I suggest the following: For the last year of your research, you must schedule two to three separate meetings with your individual committee members to go over progress and research items that you are working on. Remember that these meetings are not to take too long; 20 to 30 minutes should be plenty. Explain your progress and ask the CM if they have feedback, questions, or thoughts. Write them down and follow up on these items via e-mail.

Now here is the clincher. From 3 to 6 months before you actually defend, set up a committee meeting to go over a mock defense with highlights of your work and what your plans are for the final stages. Listen intently and take good notes when they provide suggestions because it is highly likely that they will ask those very same questions again at the final defense. This way you get to work out all the wrinkles way in advance. I've seen several students who do business this way, and I always walk away telling myself that all students should follow this model of having precommittee meetings. There is less fuss at the final hour, and the ride is much smoother.

During the last year of your research, meet at least three times with each CM individually and provide high-level results. Have a meeting with all members 3 months before defense to iron out wrinkles. If a committee member is outside your department, you could make a special trip to meet with him and present results to a larger group at his lab or university.

Oftentimes, an advisor may have the poor practice of delaying too long before letting a student wrap up their research and defend. The student can use the CMs to influence the decision-making process of the advisor. Your goal must be to make your CM work. If they are not willing to work, then they should not be on your committee.

Journal Entry

1. List your ideal committee members based upon the discussion in this book.
2. I will purpose in my heart to make my CMs work effectively.
3. I will purpose in my heart to pick the best CMs for my research—not easy ones based on subjective decisions.

Food for thought

1. List the qualities you think make an ideal CM.
2. List items that prevent you from making periodic contact with a CM. How can your circumvent them?
3. One of your CM is beginning to act strangely and all of a sudden is "too hard" on everything you do or say as far as your research goes. How should you handle such a situation?
4. If you were a CM, how would you like the student to interact with you as a CM?
5. Do the answers to questions 1 and 4 match? Why or why not?

Become a Team Player

Whose sandbox is it?

As brilliant as Michael Jordan was on a basketball court, he needed strong team players to win championships. The Chicago Bulls won when it was not just Michael who was doing all the scoring. Same thing applies in research. You have to learn how to be a team player!

Have you ever come across someone who likes to hoard information and resources to gain a competitive edge? I certainly have. It is highly likely that you will encounter someone, or several, depending upon where you work. You often wonder what makes someone that way or if it is beneficial to be closed-fisted about your work.

It is also possible that you will encounter other people who love to get, but whenever they are asked to give, whether it is information, documentation, software, or a data set, they quickly back away! I certainly have come across those people too.

Why would I start a section in such a negative fashion? Because I am hoping that you won't do either of the above! I hope you agree that we are living in an increasingly connected world. Gone are the days when you can be alone in a cave doing your own research, oblivious of what is going on around you or who is doing similar research. This is the age not only of interdisciplinary work but of multidisciplinary work as well. Read any proposal announcement from any agency. You are asked to solve a lot of things. Plus they want you to make sure that your research reaches the common person in an easy to understand format. Outreach and education—you will be expected to work in that realm as well!

Being a team player simply means that you can play nicely in a sandbox together with your colleagues with no shoving and pushing around. Mean players will be isolated and quickly given the cold shoulder. Do you ever watch kids play in a sandbox? There are some nice kids who walk around, pick up some sand, get some toys together, and then settle down to play with other kids. They talk, smile, and probably even roll around in the sand for a while. A tussle or two is often common, but the nice kids eventually get back to playing well together. Some even build a castle or a huge truck with sand and other toys. Playing nice in the research world is very much like that; a tussle or two is common between grownups, but you go back to playing nice.

One of my collaborators is seen as a tough guy to play nice with in a sandbox. He has some good sand and some great toys, but he likes to be the keeper of the sandbox, the castle, and everything, including the wood frame! I've learned that if I get into his sandbox and put in a few of my toys in a very nonthreatening way, I am able to play. Very soon, you realize that some of these tough kids are really nice, and they want to share. The trick is to get into that sandbox and play well. You have to learn how to become secure in your strengths; otherwise, you'll have a tough time becoming a team player. We actually play well together now. He still owns the sandbox (at least he thinks so), and I am a good playmate!

Another former collaborator of mine is a fantastic researcher and really knows his science. The problem with this colleague was that he never trusted anyone's skills and had a tough time developing meaningful collaborations. Sharing code was burdensome, and he only shared things that he felt wouldn't hurt him. A myopic view, wouldn't you say? Very soon, he found out that people were plenty smart to simply write the exact same code, even a lot better. They moved on without him. A tough lesson to learn and a lost golden opportunity to move forward! But enough of sandboxes and insecure scientists!

Given this backdrop, it is no wonder that team player skills are critical. The term "team player" could get very fuzzy and may lack the details that you are looking for. Here are some pointers on how you can become a team player while in graduate school. These pointers will also help pave the way for your career.

As a graduate student, you are in a fantastic position to observe how team dynamics work. You may even be lucky enough to be part of a bigger group that puts team play into action with the collective goal that everyone benefits and everyone's career moves along at a good pace. I am also hopeful that your team lead knows what it is to be a team player in a larger context. In that case, the team lead has put into place a system so that different members of the group interact with one another on a regular basis, freely share ideas, and perform quality research that acknowledges the appropriate members. Of course, while this is the ideal scenario, teamwork often breaks down because of poor work ethics, backbiting, hoarding information, and suspicion of one another's motives.

Let's talk about some practical issues. Your first mission, of course, is to do the absolute best in your classes, pass all of your exams, and then do some leading-edge research with all the principles I have talked about throughout the book. Toward the last 2 years of your graduate school career, it is often a good time to start thinking about honing team player skills. This could be difficult since you have to navigate a maze of issues such as authorships, research propriety, and a host of other questions: Will he steal my ideas? Will she share? Will I be offered coauthorship? All that fun stuff.

I'll provide some guidance. But, fully realize that everyone is trying to do the same thing and be a team player. Eventually, everything does work out, and you learn how to play well together, but some guidance does help.

As part of the taking-initiative idea, it is good for you to learn how to set your research in different contexts and sometimes within a larger context. By this, I mean that if your research can be applied to other research topics, you must take initiative to reach out to another team or another scientist and clearly explain how you would like to collaborate. There is much debate on how coauthorships should be given, if it is even meaningful to have 10 or 15 authors on a paper, and what their contributions might be. Each person listed on the coauthorship list must contribute in some form or fashion and must be given precedence on that list depending upon their level of input into that paper, but that is hardly the case in some disciplines. It is quickly becoming fashionable to simply list the first author, and then for a lack of a better system (or for offending someone) an alphabetical list based on last names follows. So if your last name starts with a Z, then you are bound to be at the end of a list. Let me rub it in—my last name starts with a C.

If you have requested collaboration or if you have obtained data or other tools from other groups or scientists, it is good practice to clearly convey what you plan on doing with this information. This should be done at the beginning of the dialog with the group or scientist. You will be expected to write papers and offer coauthorship on these papers.

It may sound strange when I say that if someone collected data you should offer them coauthorship. Sounds dodgy to you? Not really! Let me explain. Someone who collects the data is an expert in that field of interest. It is safe to assume that they know the best about that particular data set and the quality of that data. Having that person on that coauthorship list allows you to ask all the relevant questions regarding that data set, and they could also provide a section detailing their data set that could be a tremendous asset. While some data providers will involve themselves rigorously in the writing of the paper, others will simply make sure that their data have been used properly. Either way, it is only fair that you offer them coauthorship. You are probably wondering if I collect data or if I am an experimental researcher of some nature. Hardly! I do not own any piece of equipment or collect data of any kind. I believe that by including data providers in the coauthorship list and soliciting advice, it is a win-win situation for all.

The most important thing for you to do as an investigator early in your career is to show yourself accountable and diligent in pursuing these collaborations. This means that if you request data or other tools in research and if you are given these, then the pressure is definitely on you to use that information to produce good research and a high-quality research paper. If you establish yourself early on in your career with such an approach, it is fair to assume that other researchers will come looking for you to collaborate. While this could be flattering, make sure that you do not overpromise and undersell. Quality is the key!

One thing that I look for and selection committees look for in general as you grow in your career is your authorship list. If you and your advisor are the only ones on the list for a long time in your career or you are the only one on the author list writing all the papers, it signals something. Regardless of productivity, you are a loner who has trouble connecting with others. Think of it this way, you could diversify your research portfolio if you connect your research to others in your field.

You cannot expect to be successful by being a loner in this interconnected world. Beware of some minefields though. You could be so aggressive in staying connected that you could lose focus and loose integrity. If you give your word that you will write a paper using someone's data, make sure that you follow through with what you stated upfront. If you lose integrity, word gets around pretty quickly, especially in smaller research communities. This could prove to be detrimental to your research career.

Notes

Jobs

Danger of Not Finishing

The dangers of accepting a job before finishing your dissertation.

It's January of a brand new year and everything is on target for you to defend your thesis and graduate by December. The culminating moment is almost here—the proverbial light at the end of the tunnel. It is springtime, and things are still looking good for that grand finale. The trees and flowers are beginning to bloom, and you are walking a foot taller. OK, I'll get to the point!

Enter monkey wrench: a job offer. This job has come looking for you. What do you do in this case? You have a short period of time before you graduate, and a job offer has come knocking at your door. This happens too many times. Why is this tempting? First, it's the money. Let's be honest; that first real pay check would really feel good! Next, it is the job itself. This is your chance for a real job with real job responsibilities. The students who accept these offers are called ABD's—all but dissertation!

I bring up this scenario because it is quite commonplace. Remember that your end goal is to get a job. The question is this: Should you get it with or without the degree? I often ask the student to think long term in this case. While all students will say that they will finish their degree while they are working, in my opinion, this a huge mistake. It is better to finish your degree while you can. I have seen several students who take the job and either never finish or finish after several years. They had 9 months left, but it took them several years to finish. Oftentimes, the student had to relearn some of the tools to get to the point where they could actually finish their research, or they remain ABD forever.

Once you take up a full-time job, you may have to change locations, readjust, learn new things, and take on new responsibilities. There is very little time during the evenings or weekends to actually write the dissertation or finish the research. Holidays are especially hard since your mind tells you that you need to enjoy all that money you are making in that real job.

> **Bottom line:** Finish your degree and then get on with your life. You've worked way too hard to be a perennial ABD!

Staying on

Should I stay awhile after I graduate?

This is a difficult topic to handle, but it comes up a lot. I find many students coming to my office with this dilemma: "I am going to graduate next month, and I have two job offers. One job is to stay right here with my advisor and continue with my work. The other is in a certain lab/university." Ever since I have taught this course, I have become a resident psychologist of sort. Even though I have a formal master's degree in psychology, it is not in clinical psychology but in industrial/organizational psychology. Still, students tend to migrate to my office with the hope that I can provide answers. (It also has given me a lot of writing material.)

This is a tough question, and I will be as practical as possible and provide some advantages and disadvantages for each scenario. While the student usually already knows what they should do, they want a sounding board. I listen and ask questions when appropriate, and oftentimes the students will come to a conclusion during our discussion or some time thereafter. If you are an advisor reading this section, this tip is helpful. Learn how to listen to your students!

First, let me be blunt enough to say that if you have positioned yourself for the job you have desired and it has come knocking at your doorstep, then you should leave. There is no point in staying with your advisor. Now if you really are not sure

126

about this job offer for various reasons, then it may be time to sit down and draw up a pros and cons page for each job and look at them very closely. Oftentimes, students forget that their spouse should also be fully involved in this process as well.

Obviously, the biggest advantage of staying with your advisor and the team that you are currently involved with is, let's face it, familiarity. You do not have to move even an iota from the comfort zone you have developed. Everything is in place. You do not have to relocate and find another place to live. You have a social network outside and inside of your work and where you live. Better still, you have managed your advisor and figured her out. So there is no sense in starting this over, correct? Be that as it may, the comfort zone could also be detrimental to your career. The biggest disadvantage is that you will always be seen as one of the team members of your advisor's group. Now don't get me wrong; there is nothing wrong with that. But if you have always wanted to be your own person, then staying with your advisor will only make a longer transition from being a graduate student to becoming your own person.

A nice advantage of staying on is the fact that there is no loss of productivity from moving and getting set up elsewhere as there would be with a new job. However, if you decide to move, you need to prepare well in advance so that you take your loss of productivity in stride. Any move requires immense effort in all aspects of life from finding a home to adjusting to life in another city to getting car registration and grocery shopping done. So if you move, expect your productivity to take a little dip for a while, but it will come back up provided you keep your focus.

If you do stay on, then you could potentially become more productive because you are not a graduate student anymore. You have been hired as a staff member or a postdoc by your advisor. You can very quickly write more papers and proposals and increase your research portfolio. However, this should not be the only factor stopping you from taking a job in another place. You should think about your career goals and how the job fits into your final objectives. I've said it before, and I will say it again: You should think very carefully about with whom you are working if you want to move forward in your career. This is more important than where you are working. The beach may be wonderful, and the pay may be alright, but if the person that you are going to work for is an established pain in the neck, then your everyday life will take a beating.

True story! A good friend of mine accepted a job because of the comfort zone factor and the geography of the place, but he didn't do any leg work on the organization or the person he was going to work for. Very quickly, he found out that the person he was working for had no inclination, not the remotest bit, of encouraging my friend's career. In fact, every innovative idea he came up with, whether it was for a paper or a new idea or collaboration, met with stiff resistance. He was told what to do in spite of the fact that he had a Ph.D. He had to toe the party line. As brilliant as my friend was, his everyday work became drudgery, and his productivity and well-being suffered.

> **Bottom line:** Make sure you know who you will be working for, the organization and the person! It makes a huge difference.

I have several stories like this, good and bad, but the bottom line always has been this: Without vision and forward thinking, there will be much chaos later on!

Sometimes, it does happen that the place you graduated from could allow you a short holding pattern until you move on to something different. In that case, be totally upfront about it and tell your advisor that it is indeed a holding pattern until you move on. Be open enough to provide some realistic target dates. For example, "I will definitely work here as your postdoc for 1 year to 18 months. At that point, I will be moving on." This builds a good relationship between you and your advisor and provides him with enough time to recruit others for the team.

Take-home points

- Before you accept either job, take some time to write down the advantages and disadvantages of each job.
- Make sure that the job you accept fits with your career plans.
- Involve key people in your life when making decisions such as mentor, spouse, etc.
- Do a thorough analysis on the organization, the person, and the team that you will be working for.

Food for thought

1. How long do you think you should stay with your advisor after you graduate?
2. If you were the advisor, would you like to employ your student in your group after he or she graduates?

Postdoc

The waiting pattern...

There's been a lot of discussion in various books and papers on who a postdoc is and what the responsibilities are. Let me get into trouble by saying this—it is the truth though—no one really likes to be a postdoc. Let me explain why. First, a postdoc position in any university, lab, or other organization is a temporary position. It is usually a 2 to 3 year position that is specific to a certain project. Why so? Because either someone wrote a proposal and wants some short-term help on a project, or some specific expertise is required. Granted that postdoc position could become more permanent in nature, but at the onset, it is tempo-rary. That's why no one likes it!

However, postdoc positions are not necessarily a bad thing. They could serve as a very useful holding pattern until the appropriate job comes along. Here are some practical suggestions for navigating a postdoctoral tenure:

- Resolve in your mind that this position is a way to prove yourself and build a solid portfolio for the more permanent job that you have in mind.
- Even before you start your job, make specific plans on what you will do in this position to enhance visibility for you and your research.
- If you aspire to a faculty position, then make sure that your postdoc position allows you to write and manage proposals.
- Write a number of high-quality papers.
- Develop a Web site to advertise your work.
- Volunteer to develop a class or special topics either at your university or a nearby university. Universities are looking for specialists to teach courses that will broaden their students' portfolios.

- Travel to conferences but have a purpose-driven mindset.
- Make sure that you work with your supervisor and show yourself diligent in finishing that project. Your number one obligation is to satisfy the project requirements for which you are hired.
- Take ownership of the postdoctoral project. After all, your supervisor will be writing a letter of recommendation for your future job.
- Develop solid team player and communication skills in this position.
- Volunteer to mentor someone in their career while navigating yours.
- Learn how to broaden and strengthen research skills.
- Make sure that you vocalize your plans to your supervisor so they clearly understand your exit strategy.
- Finally, make sure that you are working with a mentor who can guide you through your postdoctoral tenure.

Get a Job

Every worker must be paid due wages.

Unless you are independently wealthy and you just went through those exams, writing, and pain for fun, I am assuming that your goal is to get a job after you graduate. You obviously have a desire to take your position in this world and have a career. Well, you've come to the right section then.

If you have done all the things that I have recommended in this book—managed time, reduced stress factors, focused on work, taken ownership of your research, published extensively, read papers, made sure that you are current in your field of research, traveled and presented at conferences, networked adequately, written proposals, and positioned yourself for success—then you should be the one in the driver's seat. The employer already considers you a "rising star" and will come looking for you.

In that case, you should have several options. By this time, the community knows when you are expected to graduate, and already the dialog between your advisor and the employers have started. Several employers want you to come and join their team. As I was wordsmithing this section, one of the rising stars in our department stopped by to let me know that, after deliberating on several job offers, he had decided to go to the East Coast to work at a prestigious national lab. Obviously, this stuff works!

While this is the position that I have tried to get you to in this book, the choices are sometimes daunting. Your advisor also wants to keep you in your team as a postdoc or in a staff position of some sort. You are torn between the choices as well as the "do I stay or do I leave" dilemma.

It is next to impossible to provide concrete recommendations as to what you should do when you find yourself in such a situation in a book. I have a tough time when my students who are about to graduate find themselves in my office wanting me to tell them what to do. As an advisor, it is easy to want the best for my team. Obviously, a prized student like this would increase productivity for my group. Notice that I said "my." But as a mentor, I must think about what's the best solution for the student. Keeping that responsibility in mind, I have always provided advice and allowed the student to use me a sounding board. I want the students to make that final decision. It is time for them to make real-life decisions. It's time to fly!

Having said that, there are some guiding principles that you need to follow or at least think about.

If you are the proactive, full-of-initiative type of student, then you will be totally dissatisfied if your future place of employment and your team lead (note that I did not say boss or supervisor!) have no plan in place for you to mature to independence. I have often witnessed bright, young minds get trapped in a place with no obvious way to mature and grow their careers. Frustration will set in quickly, and your productivity will suffer.

Once you start working, if you constantly dwell on things other than your work and how to increase and improve productivity, you will have a tough time maturing in your career. This is especially true when you are starting your career. This is why it is absolutely imperative that you do your homework before you start working for someone and find out all that you need to. Otherwise, you may be in for some unpleasant surprises.

True story! Take this case of a Ph.D. student who is excited to work for an organization but never inquires about the mechanics of proposal winning and project management at this organization. A few months after he starts work, he comes across a great proposal opportunity. When he approaches his team lead about writing a proposal, he is told that while he is encouraged to write a proposal, he cannot be the principal investigator and neither can he manage budgets or be fiscally responsible. Obviously, this student was devastated all because he failed to ask the right questions before he made his decision to join this team.

In this scenario, I have watched a downward spiral in the young professional because all he can think about is the issue of not being able to be a principal investigator. Productivity decreases and motivation goes down. As for job satisfaction, don't even ask!

Don't get me wrong; I am not advocating that everyone needs to be a PI. It depends upon what you want for your career. If you want to be a PI, build a team, and grow in that type of a career path, you better do your homework ahead of time and make sure that your future employer will allow for that!

There are many types of jobs and the obvious ones for someone with a graduate degree, especially a Ph.D., is academia. Even in academia, there are several pathways, including becoming a professor or starting as a postdoc or a staff member in a research scientist category. Some places allow you to be a full-time researcher and teach an occasional class or two provided you are affiliated with a university. Your title could be Research Assistant Professor. There are many that I know who absolutely do not want to be a professor anywhere, even if that were the only job available.

> **Bottom line:** Know what you want for a job and work toward that while in graduate school!

The university job

In case you have not made the connection between proposals and the way of life in research, don't worry. I'll explain. You may have heard the terminology "soft money position." Soft money simply means that you are responsible for bringing in all the money required for your salary and any team that you wish to build. And yes, you guessed it right. All of this money comes from writing proposals. Of course, we would never throw a baby graduate student to the sharks (right away at least!). By that, I mean that no one in their right mind will expect you to start bringing in your salary from the very first year you arrive in an organization. That's why you work as a postdoc on someone else's project until you can write enough proposals to start bringing in the money. I hope you are beginning to see the connections. This is the passage of science from one generation to another!

Now don't run off and start arguing about the merits and problems with such a system. Let's keep this discussion moving along. This is the way the system works currently and has for a long time, so brace yourself for it. Let the optimist in me say that if you are doing good work and publishing and networking properly, these things (such as funding) will usually fall into place.

Regardless of where you end up, each place has its own advantages and problems. If you've found a place without problems, let me know so I can apply for a job there. The bottom line is this: you have to know the environment that you will get into, ask a lot of questions during your job interview, and do plenty of homework before accepting the position.

The way you progress in an environment like this very much depends on your ability to win proposals, write papers, and develop and manage your team effectively.

The second job in a university is the Holy Grail of being a professor. While these positions are traditionally thought of as being hard to come by, I still claim that if you want to be a professor, and if you have worked toward the various aspects of becoming a professor, you can find one of these jobs sooner or later. You can even use your postdoc years wisely to further position yourself for one of these jobs. Alright, let me give you a few thoughts about the job of a professor because I started many years ago with one of those positions. You are first hired as an assistant professor (lowest man on the totem pole!). The nice thing about this is that you are guaranteed a 9 month salary. Of course, your university could shortchange you and another Ph.D. graduate by splitting one 9 month position two ways. In that case, you are called half-time faculty, and you are guaranteed— get this—only 4.5 months of salary per year. You have to bring in the remaining 7.5 months of your salary. Yes, that happens in quite a few places. The folks giving out these positions and the ones accepting them need to seriously examine their heads.

Now if you are thinking that these 9 months of salary is great—you can write a lot of papers and proposals and guide students to success—you have forgotten one small item on that list. There is a reason that they pay you that salary. Teaching! Yes, teaching! You are paid to teach and do research and serve on committees and walk on water and fly from the top of buildings unharmed! I hope you don't take some of these things literally!

As junior-level faculty, depending upon the university, you will find yourself teaching two to three courses every semester. Some departments have only a graduate program (M.S. and Ph.D.), while others are both undergraduate and graduate. Therefore, you have to sing for your supper. Oops! I mean teach for those 9 months. In this book, I am only going to outline the criteria for the job of professor; I'll make you read another book of mine that actually shows you how to climb that ladder from assistant professor to securing tenure and becoming the grand old professor quickly. Gray hair and all!

As an assistant professor, you have about 5 to 7 years to prove yourself. By that, I mean that the university has to see you as someone that they want to invest in for the long term. If that's the case, you are given tenure and a promotion to associate professor with a whopping pay increase (not really!).

Most universities have three criteria by which they evaluate a candidate. They are (in no particular order) teaching, research, and service. Now instead of me guessing the metrics for each one of these categories at your university, I strongly suggest that you sit down and actually read the faculty handbook that you were given on the first day. It actually spells out the expectations in good detail. Better still, talk to the chair of the department or your mentor.

Teaching: Generally speaking, you are required to be an effective teacher, which is usually assessed by your student ratings, your course development, and your ability to accept suggestions and improve. Make sure that you have a good game plan for teaching. If you are not genuinely interested in teaching, you shouldn't frustrate the students by teaching.

Research: If you are in a research-heavy environment, this usually means your ability to win extramural funding (fancy word for getting money from elsewhere), writing papers and winning accolades for your research is important. Remember that of each dollar you win, 50% goes to the university. Therefore, the more money you win, the harder it is for the university to reject your tenure application in a few years. Also, if you have written a sufficient number of high-quality papers on top of a good proposal, then the reviewer will find it harder to turn your proposal away. I hope that my theme of "write your papers" is making sense. They are all tied together—winning proposals, getting tenure, getting promotions, and moving ahead.

Service: This usually means that the university wants you to serve your department, college, and the university by participating in committees and staying engaged in university affairs. This means that if the chairman of your department wants you to be on the parking committee (yes, there is such a committee in most universities), you must attend those meetings, listen, and offer valued opinions on parking-related issues. This will provide you with a rounded experience that will help you in the years to come. I hear you saying, "He cannot be serious about this parking committee thing." I won't say a thing more about university committees. Wait, I am tempted. How about the committee that meets for several hours to decide on which side the faculty and the students should walk during graduation? OK, I'll quit!

Remember that many departments and universities are big into this thing called "being collegial." It means that you do not ruffle feathers, and you willingly (!) do the assigned jobs. The chair of the department or whoever runs the show knows this: The low person on the totem pole does a lot of grunt work!

Service should also include serving the research community. This involves reviewing papers for journals, reviewing proposals for agencies, and serving on national and international panels. This is probably what most younger (and older) professionals enjoy the most, unless of course parking at the local campus and where the faculty sit at commencement is important to you. Enough already!

Once you climb up the ladder and get some gray hair, you can rest a bit easier since someone else will be looking up at you from lower on that ladder. They get to be on that parking committee!

Now I hope you have an appreciation for the various facets outlined in the book: from time management, to taking initiative, to proposal and paper writing. You will be doing many more things, yes many more things when you start working. What better place to learn how to manage time and stress and put a system in place than in graduate school?

The nonacademic job

Lest I make you think that the only place you can work after obtaining a graduate degree is in a university, here are some thoughts on nonacademic jobs.

There are many places that would love to hire you now that you have all of these fantastic skills that you have accumulated over the course of your graduate career. After all, you have diligently read this book and completed the journal exercises. You have become some sort of a superperson!

You can always check out job announcements in bulletins, journals, and Web sites pertinent to your academic discipline, or if you have been making the right contacts throughout graduate school, your task of finding a job is easier. Note that this is not an exhaustive manual or book that lists all the various places where jobs are available. I am only providing some pointers on how to position yourself for success. With that in mind, here are some possible career paths.

The federal job

The federal job or a civil servant position is still coveted by many and rightfully so. It is often prestigious and pays reasonably well. As a bonus, the pressure to write proposals and secure funding is not as intense as in a "soft money" job. At least, that is the perception. If you do not want to teach but want to be a civil servant, then a federal job is a good option. These jobs are not necessarily limited only to the National Weather Service in forecasting positions. Various national labs and other organizations such as NASA, NOAA, and the Office of Naval Research employ students with good technical and communication skills. However, only U.S. citizens are eligible to apply.

How does one get a federal job? There are several options. Like the lease-to-own option, there are several fellowships available that allow you to work and go to graduate school at the same time. Once you graduate, you are placed within the organization based on the needs and requirements of that job. There are obvious advantages to this option. While there is stiff competition for these fellowships, once you win, you have your short-term career path mapped out for you. However, make sure that you read the fine print on all the requirements. The major disadvantage is that it may not be possible for you to know where the organization will place you after you graduate.

The other option is waiting until you graduate to apply for a civil service position. You have to apply formally for this position, which means that there may or may not be positions open after you graduate. It is indeed a waiting game.

What are some of the metrics for landing a federal job? It is mostly the same for any other job after you get your graduate degree. Federal positions call for good scientists who have the ability to apply their research to a larger context and—get this—communicate well. Oftentimes, these positions are looking for someone who will have the capacity to lead after someone retires from a key position within the organization. Therefore, you must show yourself approved in the same areas that we have been talking about: communication, vision, organization, and a solid science portfolio. Often, there is a miscommunication that published papers may not be necessary in this regard. I disagree. If you apply for such a job with peer-reviewed papers, you will be known for your ability to take research to completion. Whether it is operational or research-oriented science, the ability to publish papers is priceless!

The private organization

Ah! The world of the private organization—lots of money, stock options, and great retirement and health plans. Wait, before you get carried away! Jobs in the private industry could be quite rewarding provided they match your aspirations and career goals. There are a wide variety of jobs available for students with a M.S. or a Ph.D. A fast exploding area for jobs is geographic information systems (GIS). I have had a lot of discussions with colleagues around the country who are always looking for students with the right mix of technical and computer skills coupled with a knowledge of GIS.

What do private organizations want? Mostly the same set of skills that have been discussed in this book but with an even stronger emphasis on communication skills. First, the employer's needs must match the student's KSAs—knowledge, skills, and abilities? Next, the ability to take leading roles in projects could be useful. If you have trained and honed your "taking-ownership" skills, this could be useful. You can even use that phrase in your interview—taking ownership. Employers love it!

Going abroad: While most of us dream of doing this while we are graduate students (read as, "I want to go abroad to do a postdoc and then come back to the United States"), for whatever reason, only a few actually do this. There are numerous reasons why one would do this. These reasons range from a potential life partner forcing you to go abroad to simply wanting to spread your wings a bit. Whatever your reason, the underlying principles are the same. If you want to get back to the American system after a stint (no, I did not say frolic!) abroad, then you have to think ahead and do the necessary things that will keep you positioned for your reentry into this system. It will be difficult for you to gain good employment in the United States if you were unproductive (no peer-reviewed papers in major journals) in your years away from home. This is largely because the American system thrives on doing competitive research in most if not all areas. However, you could be seen as an asset if you have developed appropriate networking skills and if you show the pathways for collaboration between your home country and your place of employment abroad. While at it, learn to speak a foreign language!

While these are the most common routes from graduate student to a salaried job, there are other possibilities as well. I have made the assumption that you have entered graduate school with the intention that you want to contribute to your field while enjoying what you do and making some money along the way. I've tried to present some of the issues related to each job. No doubt, there are numerous other issues that you will become aware of that you wished I had written about. Every organization is unique, and there are bound to be differences from the norm. Otherwise, life would be boring. Learning how to navigate these differences while keeping your career goals in mind is important.

Here are some interesting (aren't they all supposed to be interesting?) thoughts on this section. It is good for you to look at all these options before making a decision on where to start working. The primary objective is to know what you want for your career in the near term, midterm, and long term. If you cannot think long term, it is quite alright. When you get to midterm, you will know. If it is any consolation, I have no clue about my long-term plans, and I am already at the midway point of my career (I think!). If you have no idea as to what you want even for the midterm, let me at least point you in the right direction so you won't make huge mistakes in your job selection. I just got a serious talking to from one of my mentors about developing a long-range plan. Groan!

The organization that you will join should not be an isolated place. By this, I don't mean that this is a high-latitude location and geographically remote. Just to be sure, I'll say this: A lot of geographically remote and frigid places have excellent research programs. If the place of your employment has very little by way of resources (computers and other things vital for your research, for example) and if you do not have anyone to interact with in your area of expertise, then it is a long struggle to build everything from scratch. I am not implying that you cannot be successful, but you had better make sure that you have that pioneering spirit, energy, and time to persevere through the tough times before things get better! Often, like starting a small business, a lot of physical and mental resources are necessary to start something from scratch.

If you will be working on someone's project, you need to make sure that the person has your interests in mind also. Will they provide an environment for you to develop and mature, or will they simply be focused only on their research program? If it is the latter, then this is a dangerous red flag that you could become disappointed and disillusioned along the way.

Is the team lead still actively engaged in the field or are they thinking about hanging it all up soon? If you are going to work for someone, I strongly suggest that you check out their recent publication record, their projects, and their level of involvement in the field. Carefully listen to their comments. If this person is disillusioned with the system and they are cynical about everything, then it is very likely that you will succumb to that as well. Stay away from such people and organizations.

Stay away from controlling environments that do not have a vision. They may be doing good work, but it is their good work. If they are not willing to think about your career goals, then you must decide if you can handle such an environment. I will mention this in the job interview section, but you must talk and spend some time with the team members and ask some probing questions to get a sense of the team environment that is prevalent. How accessible is the team lead? How nurturing is the team lead? How well does he manage time and stress? Are the expectations realistic?

Job Interview

How to blow away the selection committee and the competition.

As much as the organization wants you to work for them because of your star qualities, they are still bound by law to be an equal opportunity affirmative action employer. Rightfully so! Since they are so impressed with your credentials, the job announcement/advertisement looks like it was written for you. Nevertheless, a selection committee is formed, and all the applicants are evaluated. It is possible that for one position there are about 25 applicants. The committee usually creates a short list of the best candidates they feel will fit the job requirements. Obviously, you are one of the candidates.

The interview should serve several purposes. From your perspective, this is the opportunity for you to, as they say, "blow their socks off" and ask probing questions during your visit. From their perspective, they want to make absolutely sure that you are the right candidate and the right long-term investment.

Typically, you will be asked to give a 30 to 40 minute talk with 15 minutes for questions. You will also meet various people depending upon the job. If it is for a faculty position, you will meet the faculty individually plus the chair, dean, the selection committee, and, of course, the students. If the position is for a private organization, you could meet some members of the team, the team lead, and one or two people from higher up on the hierarchical chain. For a federal job, you would meet your group members, the selection committee, and some of the chain of command as well. In any case, one of the centerpieces of this visit is probably the talk that you will give.

Before you go, you must do some thorough homework so you can present your seminar appropriately and ask the necessary questions to make a decision about this job. After all, if they do make a job offer, then you must be prepared. First, find out about the breadth and depth of expertise in the organization and the members of the organization. Find out about the relevant research topics that your current and future work may apply to. If you know someone within the organization, it is not inappropriate to ask questions regarding the workplace before you go.

It is standard practice that you are e-mailed an itinerary of your visit detailing who you will meet on what days and hours during the interview process. This gives you plenty of time to prepare questions ahead of time that you can ask during your one-on-one discussions. The person who will most probably interact with you is the chair of the selection committee. To save embarrassment and awkwardness, go ahead and ask who you should talk to about salary-related issues. This way you are not left guessing, and some of the stress is avoided. Usually, you will get a straight answer, and you will know who you can talk to about this. We'll talk about salary negotiations later.

Now what type of a talk should you prepare? It is often tempting to give the same talk that you gave for your thesis/dissertation defense. After all, you have rehearsed it and passed the exam. You feel comfortable. However, in this case, I believe that this is the worst talk to give. It is too specific, and you will quickly cause boredom. Remember that your employer wants to assess your potential. A dissertation talk will be too narrow and too focused. Also remember that everyone that you will meet later in that day or those you have already met in the chain of command and the team will be there. You must develop a talk that tells the audience that you understand the broader picture of your science and the related fields. For example, if your dissertation was about developing an algorithm to calculate sea surface temperature from satellite data, it could be downright boring if that is all you were to present. A sense of why this is important to related fields and how future sensors on satellites could benefit from the research will be useful. Also, if you connect your work to others in the organization, marine sciences, ocean productivity, etc., it tells the audience that you know how to take the dots and start making connections to form a bigger picture.

The big picture is what you should strive for. Make sure that you explain some of the fundamentals of your science, the state of the art, your contribution, and your plan for your work over the next several years. If you also mention which agencies and programs might be interested in your work, the selection committee will also know that you have thought the process through. I am absolutely sure that I would be impressed with such a seminar if I were in the audience, especially as part of the selection committee.

Now I hope you are fully convinced that my section on learning how to communicate effectively while in graduate school is making sense! In your one-one-one discussions with members of the organization, you probably will get a perspective from each member, and they will have questions for you ranging from "Why do you want to work here?" to "What are your strengths and weakness?" to bizarre questions such as "How were you able to publish so many papers even before you graduated?" Smile and don't get too cocky! My advice is to be upfront and concise and show yourself to be a solid team player.

Make sure that you can ask some questions too. Ask them what they like and dislike about the organization? Ask them about how they feel you can progress in your career. Ask them about opportunities for growth and threats to progress. Ask any question you deem appropriate to help you in the decision-making process.

I have included a checklist of important questions.

- For the selection committee or others
  - What types of computing resources are available for your work?
  - What lab resources are available if needed?
  - How long is the contract and how often is it renewed?
  - What metrics are used to measure performance?
  - What are their expectations for you?
  - How are assessments and promotions handled?
- For the person in charge of fiscal decisions
  - What type of salary can be expected?
  - What is included in the start-up package?

This is standard stuff, but you do need to know what the various benefits are when joining the organization. You will be excited at the prospect of starting your career, but it is best to ask questions about retirement, benefits, and health, life, and other insurance packages. It is rare that an organization will negotiate on such standard packages. So as part of your interview, ask to speak to a human resources representative and become informed about the various packages.

- For the person from Human Resources
  - How much vacation leave do they offer?
  - What is the policy on sick leave?
  - How do vacation and sick leave accumulate over the year? If I do not use the leave, do I lose it?
  - What is the 401K or 403B package (retirement and savings packages) that you have? Do you match employee contributions? What are the limits to this tax-deferred savings plan?
  - What type of health insurance is offered? Can family be included in the coverage?
  - What is the policy for leave of absence regarding maternity-related issues?

Salary and start-up packages

Needless to say, if you have a solid portfolio of papers, research experience, leadership, vision, and the right communication skills (as they witnessed from your talk), then you are in a very good position to negotiate a good start-up package.

> **Bottom line:** The stronger the candidate's resume/CV, the more likely that the salary and the start-up package can be favorably negotiated.

Salary discussions and negotiations take many shapes and forms. Although it is tough to predict how yours might turn out, there are indeed some guiding principles. First, remember that there is a certain salary range for your position. The head of the department or the group must make a decision that does not upset the existing employees in that organization. You cannot expect to be paid 1.5 times more than the person who has been working for 7 years. You would not want that if you were the person with 7 year experience. Therefore, knowing that range (oftentimes it is hard to know) or limiting constraints for negotiating upward is important. When I began my career, the person responsible for discussing salaries did not even give me an opportunity to discuss anything. He wrapped up the discussion by noting that the salary had been set by the organization and that was that. Only a year later did I realize that I was cheated out of a good starting salary!

This is another reason that I teach this class and for writing this book! Since I do not want you to suffer the same plight, I want you to be fully knowledgeable about some of these issues. Remember that you have worked hard to position yourself for your career. Don't let someone cheat you out of a good salary. After all, each person should be paid due wages!

Next, you need to discuss start-up packages. Most places will offer a start up package of either a certain dollar value or other resources. For example, a certain sum of money is set up in an account that you could use to purchase computers, hire graduate students, or fund legitimate research needs. Use these wisely and negotiate this amount as well. Sit down and do the math of how much will cost you to have graduate students and computers and other resources that you will need to be successful. Be upfront and explain that these are some of the things that you will need to keep up your productivity. Make sure you mention that this is important for you to write successful proposals. Also, keep in mind that you need to be reasonable with what you are asking. Organizations want incoming candidates to be successful but essentially low maintenance, so when you join them, you can run be up and running from day one. Tell them that you want to minimize lost time and keep your research productivity up, so you will need such resources. Everyone loves a productive worker!

Also, ask about moving and relocation costs. Be careful on how you accept this. If you accept this as a bulk amount, you may be taxed. So carefully check out tax laws. You can also ask to be paid for actual relocation expenses that may not involve taxation. Either way, consider yourself warned.

Notes

Career Exercises

Here are a few exercises that will be well worth your time. They will help you to focus on your career goals and position yourself in a successful career.

Job Hunting and CV Exercises

- Assume that you are job hunting. Look for jobs using any method that you choose and write a cover letter with a complete CV.
- Using the job announcement as a guide, conduct a job-specific SWOT analysis.
- Determine how you will mitigate your weaknesses so that you can become a top candidate for these jobs.

Proposal Writing Exercises

Assume that you have just finished your Ph.D. Your task is to write a new investigator-type proposal. Pick any agency of your choice that lines up with your type of research. Look for the young or new investigator announcements. Write a full proposal following all guidelines.

Notes

12

Concluding Thoughts: The Final Word

"Really?"

I feel like I have provided tons of two-word phrases such as write papers, manage time, avoid stress, take initiative, manage advisor, take ownership, stay proactive, and the list continues. But, it is written with you and your career in mind. I want you to position yourself now for a successful career. Without vision, there is bound to be dejection, disappointment, confusion, and outright chaos!

This is my first foray into book writing, and no one ever knows how things will pan out, especially for a book of this kind. As I said in the introduction, it is tough to capture the classroom dynamics in a course of this kind. I have tried to be upfront about career issues in this book. More importantly, I hope I have provided valuable information on what you should be doing now for a successful career after you have graduated. I hope that as you have read this book, some of the fears about writing papers and proposals and navigating job interviews have worn off. I also hope that, for the first time, you have actually learned how to manage your advisor and what he or she expects from you. You can be assured that, if all goes well, there will be another book written for the advisor. There is bound to be a chapter called "How to Manage Your Student"!

Jokes aside, most people face pressures related to time, and stress becomes inevitable. I have tried to provide a student perspective on how to mitigate some of these issues. *If you get one thing out if this book this should be it: Sow well now to reap big later!* It is like saving for retirement. If you want to retire, you had better save money. Everything that you do in graduate school should be with the express purpose of making the path easy for yourself at a later stage. Sow now to reap later, and sow on good ground!

Think of your graduate work both in broadly based and in compartmentalized terms. You need to know where you are going and learn how to break it down into small segments so that you learn to experience success along the way. Build a strong foundation on which you can stand, learn the tools of your trade well, and communicate effectively. There is no place like graduate school to learn all of this!

I have emphasized paper writing quite a bit (a lot!) in this book because I believe that it provides needed focus for the student. After your foundation is built through classes, then you must write papers—a lot of papers. Writing papers requires you to read a lot of papers as well. Focus on the important, compart-mentalize research, and, most importantly, experience success. Don't get too narrowly focused too soon in research. Try some diverse ideas before you get down to your actual thesis or dissertation topic!

Remember that your advisor wants you to succeed at all levels. They are your allies not your foes! They know that if you do not experience success as a student while in graduate school, disillusionment could set in.

Finally, take ownership of your research. You will be satisfied with this approach, and all the pains of graduate school will seem to fade away.

There are really no limits unless you impose them on yourself.

Best wishes for a brilliant career, and I look forward to hearing from you!

Notes

Index